Hydrogen Production for a Hydrogen Economy

Topical Conference at the 2008 AIChE Annual Meeting

Philadelphia, Pennsylvania
16 - 21 November 2008

ISBN: 978-1-61567-205-9

Printed from e-media with permission by:

Curran Associates, Inc.
57 Morehouse Lane
Red Hook, NY 12571

Printed by Curran Associates, Inc. (2009)

For permission requests, please contact AIChE
at the address below.

AIChE
3 Park Avenue
New York, NY 10016-5991

Phone: (203) 702-7660
Fax: (203) 775-5177

www.aiche.org

Additional copies of this publication are available from:

Curran Associates, Inc.
57 Morehouse Lane
Red Hook, NY 12571 USA
Phone: 845-758-0400
Fax: 845-758-2634
Email: curran@proceedings.com
Web: www.proceedings.com

TABLE OF CONTENTS

Novel organic solvents for the Bunsen Reaction

Marie L. Taylor[1], Peter Styring[1] and Ray W. K. Allen[1]
[1]Department of Chemical and Process Engineering, University of Sheffield, Mappin Street, Sheffield, S1 3JD, United Kingdom

Introduction

Hydrogen is an energy carrier that has the potential to be produced from renewable sources, with minimal (potentially zero) harmful emissions to the environment. There are many technologies (both proven and non-proven) for the production of hydrogen. Thermochemical cycles linked with nuclear heat are one of the most promising routes [Ewan and Allen 2005]. Of these cycles, the sulphur iodine (SI) cycle appears to have encouraged the most international excitement over the last decade.

The SI cycle was first developed by General Atomics (GA) in the early 70s. The cycle is characterised by the following reactions:

Equation 1: $I_2 + SO_2 + 2H_2O \rightarrow H_2SO_4 + 2HI$

Equation 2: $H_2SO_4 \rightarrow \frac{1}{2}O_2 + SO_2 + H_2O$

Equation 3: $2HI \rightarrow H_2 + I_2$

Equation 1 is usually known as the Bunsen reaction. Here iodine reacts with sulphur dioxide and water to produce two acids, sulphuric acid (H_2SO_4) and hydriodic acid (HI). These acids are then decomposed, the H_2SO_4 decomposition produces an oxygen product (equation 2) and the HI decomposition produces hydrogen (equation 3). The overall result is a net input of water and a net output of hydrogen and oxygen. All other components are recycled.

For the Bunsen reaction, GA put forward a processing scheme where all reactants and products are fluids. The melting temperature of iodine is 387K so the Bunsen reaction was carried out at 393K. At this temperature the reaction is unfavourable, the free energy change being +82 kJ mol^{-1} (400K). GA showed that the two acid products formed in the Bunsen reaction could be separated by density if sufficient excess iodine was present [Norman et al. 1978]. This excess iodine helps to shift the reaction equilibrium forward and also causes the acids to spontaneously separate into two aqueous phases; a light phase containing the sulphuric acid and a heavy phase containing polyhydriodic acids, iodine, water and a number of protonated polyiodine complexes (known as the HI_x phase). The formation of these polyhydriodic acids in which iodide anions are solvated by molecular diiodine causes the phase separation to occur. The phase separation characteristics of the acids increase with increasing iodine concentration up until the iodine saturation point is reached [Sakurai et al. 2000].

An excess of water is also required for the reaction in order to shift the reaction equilibrium to the right, however this causes a significant change in the enthalpy due to acid dilution. These excesses of both water and iodine lead to a modified Bunsen reaction with a free energy change of -88 kJ mol^{-1} at 400K:

$$9I_2 + 16H_2O + SO_2 \rightarrow (2HI + 8I_2 + 10H_2O) + (H_2SO_4 + 4H_2O)$$

This modified reaction becomes thermodynamically favourable but is also very exothermic (ΔH = -90 kJ mol^{-1}) due to the large excess of water required. A further disadvantage is that this modified reaction causes complications with the HI separation and decomposition section due to the presence of a complex azeotrope.

The reduction of the large excesses of both water and iodine needed for phase separation is one of the key research challenges for future development of the SI cycle. Although promising results have been achieved using an electrochemical membrane reactor for the Bunsen reaction [Nomura et al. 2005] there is a problem when scaling up to commercial size. Therefore, alternative ways of reducing the large reactant excesses are currently under investigation. A small amount of work has been conducted in the past on the use of excess liquid sulphur dioxide and organic solvents in the Bunsen reaction, as a means of complexing and extracting the HI [De Beni et al. 1980]. The authors identified that liquid sulphur dioxide, being immiscible with sulphuric acid at low temperatures, dissolved HI well and therefore enabled acid separation. Both low boiling and high boiling molecular organic solvents were then employed as materials that were miscible with the sulphur dioxide and were capable of extracting the HI. Good results were achieved with the addition of high boiling organic solvents, in particular with tributyl phosphate (TBP). However, problems arose in then separating the HI from the solvent following water removal. On heating past 350K, the TBP began to degrade.

The aim of this work is to identify and examine possible solvents or groups of solvents that could be used effectively in the Bunsen reaction. Initial work on one potential group of solvents, ionic liquids, is reported here.

Ionic liquids

In the last decade ionic liquids have emerged as excellent solvents in the area of green chemistry [Holbrey and Seddon 1999]. Ionic liquids are 'non-volatile, non-flammable and thermally stable solvents and as such are very promising replacements for the traditional volatile organic solvents' [Huddleston et al. 2001]. They are liquid at room temperature and are composed entirely of ions. The physical and chemical properties of these ionic liquids can be tailored by the choice of cation and/or anion [Brennecke and Maginn 2001] meaning that for any given application an optimum ionic liquid exists.

Ionic liquids are starting to be used in industrial separation methods and this is mainly due to their extremely low vapour pressure and thermal stability [Berthod et al. 2005]. Investigations have looked at the separation of metal ions, the separation of organic compounds, the separation of gases and the use of ionic liquids in analytical separation.

Ionic liquids have also been used in the field of acid extraction. Martak and Schlosser [2007] investigated the extraction of lactic acid with the ionic liquid trihexyltetradecylphosphonium bis(2,4,4-trimethylpentyl)phosphinate. The authors found that the ionic liquid was successful at extracting undissociated molecules of acid via a hydrogen bonding mechanism.

The work reported here focuses on whether the novel properties of ionic liquids can be utilised in the SI cycle as a solvent capable of extracting the HI into a second phase in the Bunsen reaction. For this to happen, major property requirements of the ionic liquid are that it must:

- Be immiscible with water
- Have a high selectivity for extracting HI
- Have a high thermal degradation temperature
- Have a high capacity for dissolving both iodine and sulphur dioxide

One of the first objectives of the experimental program was to identify those ionic liquids that are selective for HI via simple liquid-liquid extraction tests.

Experimental

Following a literature study and preliminary testing, nine ionic liquids were selected for further examination. In order to make a direct comparison to the work done by De Beni et al. [1980] TBP was also analysed for its extraction capability in the same environment and the results are presented with those of the ionic liquids.

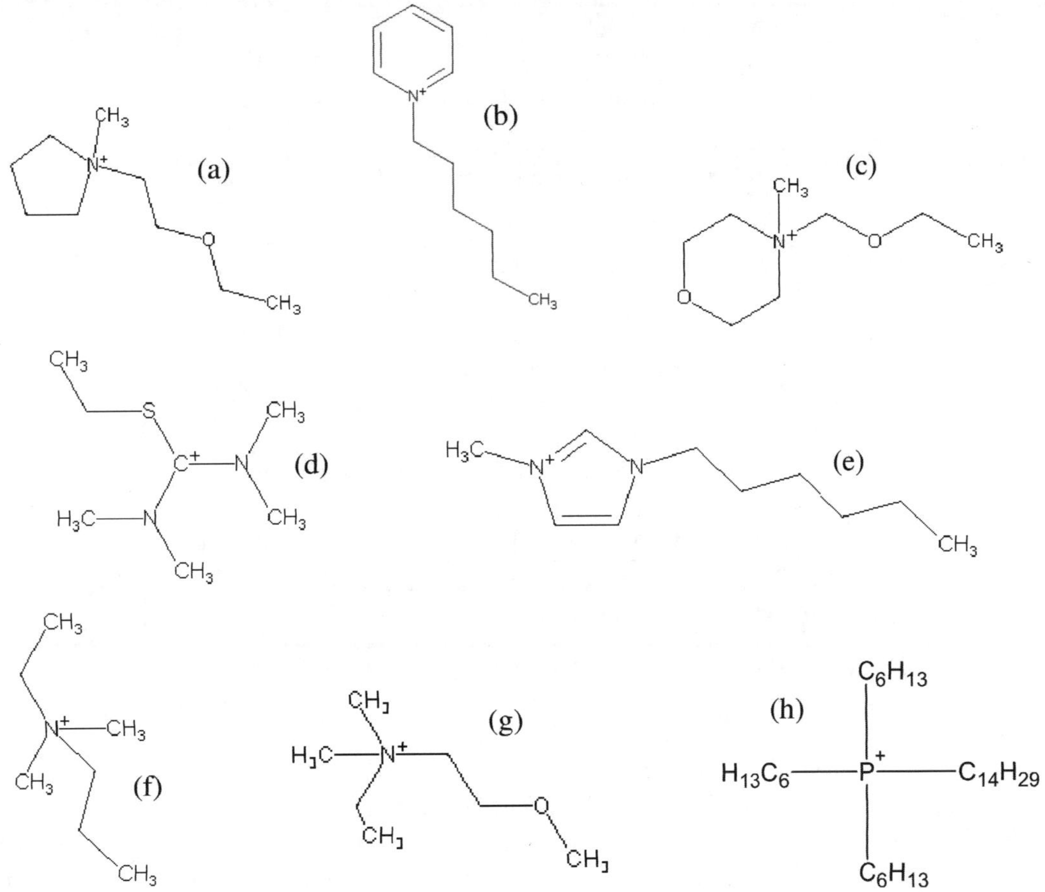

Figure 1. Ionic liquid cations: (a) 1-(2-ethoxyethyl)-1-methylpyrrolidinium ((eOe)mpl); (b) n-hexylpyridinium (hpyr); (c) N-Ethoxymethyl-N-methylmorpholinium ((eOm)mmo); (d) S-Ethyl-N,N,N',N'-tetramethylisothiouronium (Sem_4iso); (e) 1-hexyl-3-methylimidazolium (hmim); (f) Ethyl-dimethyl-propylammonium (Nemmp); (g) N-ethyl-N,N-dimethyl-2-methoxyethylammonium (Nemm(mOe)); (h) trihexyltetradecylphosphonium ($P_{666,14}$)

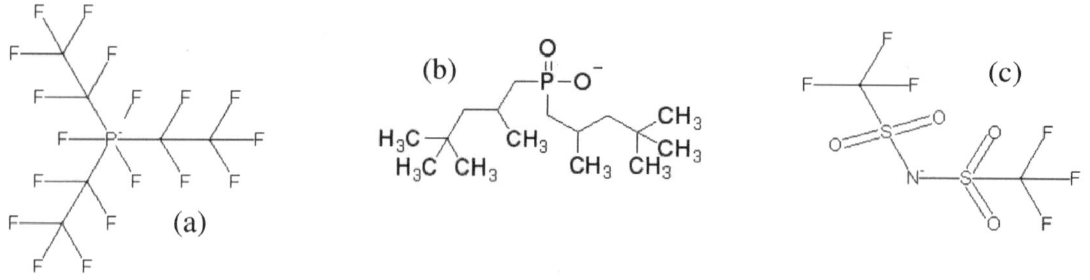

Figure 2. Ionic liquid anions: (a) tris(pentafluoroethyl)trifluorophosphate (FAP); (b) bis(2,4,4-trimethylpentylphosphinate (TMPP); (c) bis(trifluoromethylsulphonyl)imide (Tf_2N)

Figures 1 and 2 show the molecular structure of a selection of ionic liquid cations and anions. The ionic liquids used in the experiment are listed in table 1. All ionic liquids except the n-hexylpyridinium bis(trifluoromethylsulphonyl)imide were available commercially and were used as received. The n-hexylpyridinium bis(trifluoromethylsulphonyl)imide was synthesised and verified by ^{1}H NMR prior to use.

Table 1. Ionic liquid selected for HI extraction tests

	Ionic liquid
1	1-hexyl-3-methylimidazolium tris(pentafluoroethyl)trifluorophosphate [hmim][FAP]
2	1-(2-Ethoxyethyl)-1-methylpyrrolidinium bis(trifluoromethylsulphonyl)imide [(eOe)mpl][Tf_2N]
3	Trihexyl(tetradecyl)phosphonium tris(pentafluoroethyl)trifluorophosphate [$P_{666,14}$][FAP]
4	N-Ethoxymethyl-N-methylmorpholinium bis(trifluoromethylsulphonyl)imide [(eOm)mmo][Tf_2N]
6	S-Ethyl-N,N,N',N'-tetramethylisothiouronium tris(pentafluoroethyl)trifluorophosphate [Sem_4iso][FAP]
8	Ethyl-dimethyl-propylammonium bis(trifluoromethylsulphonyl)imide [Nemmp][Tf_2N]
9	N-Ethyl-N,N-dimethyl-2-methoxyethylammonium bis(trifluoromethylsulphonyl)imide [Nemm(mOe)][Tf_2N]
11	Trihexyl(tetradecyl)phosphonium bis(2,4,4-trimethylpentyl)phosphinate [$P_{666,14}$][TMPP]
14	n-hexylpyridinium bis(trifluoromethylsulphonyl)imide [hpyr][Tf_2N]

Equal volumes of each of the ionic liquids were mixed with unstabilised azeotropic HI solution for one hour. Following this the mixtures were left to settle for a minimum of 12 hours. Samples of both the aqueous and organic phases were analysed for iodine, acid and iodide content using potentiometric titrations. An iodide balance and a proton balance allowed independent determination of the mass of both phases. ^{1}H NMR spectra were obtained for each of the pure ionic liquids and also for the aqueous and organic phases following extraction. This allowed any chemical interaction between the acid and the ionic liquid to be observed.

Results – HI extraction

Table 2. Amount of protons and iodide (I^-) ions extracted by the ionic liquids

Ionic liquid	% Protons extracted into organic phase ± 0.3	% Iodide ions extracted into organic phase ± 0.3
1 [hmim][FAP]	0.3	0.5
2 [(eOe)mpl][Tf_2N]	22.5	11.7
3 [$P_{666,14}$][FAP]	0.4	0.7
4 [(eOm)mmo][Tf_2N]	26.5	15.2
6 [Sem_4iso][FAP]	1.3	1.5
8 [Nemmp][Tf_2N]	16.7	7.2
9 [Nemm(mOe)][Tf_2N]	21.2	9.8
11 [$P_{666,14}$][TMPP] - **run 1** **11** [$P_{666,14}$][TMPP] - **run 2**	18.0 14.7	19.3 23.6
14 [hpyr][Tf_2N]	10.8	4.7
TBP	-	15.9

Table 2 shows the % protons and % iodide ions extracted by each of the ionic liquids. The data is presented based on ions rather than molecular HI so as to show the significant difference in the amounts of protons and iodide ions extracted by the Tf_2N ionic liquids. The TBP was found to extract 15.9 wt% of the HI in the same environment. In all cases (except that of IL 11) the mass balances based on the proton and the iodide agreed to within 5%. Looking at the results, it is clear that the ionic liquids can be split into three general groups for analysis according to their anions:

Tris(pentafluoroethyl)trifluorophosphate (FAP) ionic liquids (ILs 1, 3 and 6) - The extraction of protons and iodide ions for the FAP ionic liquids show that very little of the HI from the aqueous phase is extracted when compared to that of the TBP result. It is hypothesised that this may be due to the general mechanism of extraction of most hydrohalic acids. Many authors have documented the extraction of hydrochloric acid by TBP, and with it the co-extraction of water molecules. Kertes [1960] demonstrated the presence of two well defined species in the TBP-HCl-H_2O system, the first forming at low initial acid concentrations with the composition $[(TBP)_2{\cdot}HCl{\cdot}(H_2O)_6]$ and the second forming at higher initial acid concentrations with the composition $[TBP{\cdot}HCl{\cdot}(H2O)_3]$. Stoyanov [1993] goes on to describe the molecular state of these species at different initial acid concentrations. Baldwin et al. [1959] state that for HI, hydrobromic acid and hydrochloric acid water is co-extracted into TBP at an approximate ratio of four water molecules per acid molecule. The co-extraction of water with hydrohalic acids has been observed with other organic solvents such as trialkyl phosphine oxides [Mrnka et al. 1974, Alguacil and Lopez 1996] and methyl isobutyl ketone [Widmer 1970]. During the work done on the Bunsen reaction with solvents De Beni et al. [1980] observed co-extraction of water. The authors attributed this to the strength of the HI aqueous azeotrope.

Water co-extraction has also been observed when extracting acids into ionic liquids [Martak and Schlosser 2007]. It is therefore highly likely that the extraction of HI into ionic liquids will involve the co-extraction of water.

The anion is generally used to control the water miscibility of the ionic liquid (although the cation can also influence the hydrophobicity) [Huddleston et al. 2001]. The water uptake of a selection of ionic liquids containing the three anions under investigation is shown in table 4:

Table 4. Comparison of water uptake by ionic liquids containing Tf_2N, FAP and TMPP anions. [a] Ignat'ev et al. [2005] [b] Cyphos 104 data sheet

Ionic liquid	Water uptake (ppm)
1-hexyl-3-methylimidazolium Tf2N	10670[a]
1-hexyl-3-methylimidazolium FAP	2030[a]
1-butyl-3-methylpyrrolidinium Tf2N	14800[a]
1-butyl-3-methylpyrrolidinium FAP	3500[a]
Trihexyl(tetradecylphosphonium) TMPP	206000[b]

It is clear that the order of hydrophobicity for the anions is FAP > Tf_2N > TMPP. Of the three groups, the FAP ionic liquids extract significantly less HI than the Tf_2N and TMPP ionic liquids. If HI is only extracted in a hydrated form then it would be expected that ionic liquids containing FAP anions would extract the least amount out of the three. Unfortunately, the water content of the organic phase could not be measured directly using Karl Fisher analysis due to the presence of iodine.

Although the mechanism of extraction is not fully understood, it is clear that only negligible amounts of acid are taken up by the organic phase meaning that they are unsuitable for use in the Bunsen reaction.

Bis(trifluoromethylsulphonyl)imide (Tf_2N) ionic liquids (ILs 2, 4, 8, 9 and 14)- The results for this group of ionic liquids are slightly more complex. There are significant differences in the acid content values depending on whether it is based on the proton or the iodide analysis. In each case only approximately half the amount of iodide ions, when compared to protons, are extracted into the organic phase. The mechanism of extraction with regard to the Tf_2N ionic liquids appears to differ to that of the FAP ionic liquids (water co-extraction aside). Looking at the organic phases, it can be seen that more protons are present than corresponding anions and vice versa in the aqueous phase. In order that electroneutrality be maintained throughout the system this must mean that there is transfer of ionic liquid ions between the phases. As the HI is extracted into the organic phase in its dissociated form the protons are free to pair with either the iodide ions or the anions of the ionic liquid. Similarly the iodide ions can pair with the protons or the ionic liquid cations. Of the Tf_2N ionic liquids tested none of the cations are hydrophobic in nature. Therefore, if a pairing between an iodide ion and an ionic liquid cation occurs it is free to transfer into the aqueous phase.

This ion exchange mechanism has been observed previously. Work has been undertaken to examine the extraction of metal ions from acidic nitrate media by ionic liquids [Dietz and Dzielawa 2001]. The authors observed that the extraction of strontium from aqueous nitrate media, did not involve nitrate ion coextraction, indicating a cation exchange mechanism taking place. The effect of increasing the hydrophobicity of the cation was investigated [Dietz et al. 2003] and a decrease in cation exchange was observed when the length of the alkyl chain on the cation was increased from C5 to C10. However, further work by the authours revealed that multiple mechanisms can take place, and that ionic liquid loss to the aqueous phase can still occur even with these more hydrophobic cations [Dietz and Stepinski 2005].

In summary, it is expected that the Tf_2N ionic liquids also extract HI with a co-extraction of water. Due to their more hydrophilic nature (when compared to the FAP ionic liquids) they are able to extract more acid and this is seen in the extraction results. However, as the system is entirely ionic, an equilibrium exists in the organic phase between HI, [H][Tf_2N], [Cation][I] and [Cation][Tf_2N], of which the [Cation][I] is free to transfer into the aqueous phase. Observation of the ^1H NMR spectra of the aqueous phases following extraction with the Tf_2N ionic liquids does indeed show the presence of the organic cation in all cases.

Obviously, a loss of the cation to the aqueous phase severely compromises the integrity of the solvent. Unless a process can be found whereby there is no loss of the solvent it is unliklely that the Tf_2N solvents will be suitable for the Bunsen reaction.

Bis(2,4,4-trimethylpentylphosphinate) (TMPP) ionic liquids (IL 11) - The results for the bis(2,4,4-trimethylpentyl)phosphinate ionic liquid tested show discrepancies in the component balances. The potentiometric analysis of the organic phase for both protons and iodide ions exhibit two equivalence points, suggesting that there is a more complex chemical interaction occurring than seen with the previous two groups of ionic liquids. Further work is needed to determine what is happening in this mixture, however significant amounts of both protons and iodide ions are extracted when compared to the TBP.

Conclusions

The results presented in this paper show that a great deal of further work is needed if ionic liquids are to be considered for use in the Bunsen reaction of the sulphur iodine cycle. Unlike more traditional molecular solvents, the extraction mechanism is potentially far more complex as the entire system, both solvent and acid, is ionic by nature. It is hypothesised that the extraction of HI by these solvents involves the co-extraction of water molecules, most probably due to the nature and strength of the HI aqueous azeotrope. The extremely hydrophobic nature of the FAP ionic liquids means that negligible amounts of acid are taken into the organic phase therefore making them unsuitable. The Tf_2N ionic liquids, whilst extracting significant quantities of acid succumb to a cation exchange process, leading to solvent loss to the aqueous phase. The TMPP ionic liquid shows some promise however further work is needed to determine the complex processes taking place in the mixture. Current investigations are now focussing on more traditional molecular solvents with the aim of avoiding the cation exchange process which occurs in some ionic liquids.

Acknowledgements

The author would like to thank Professor Ken Seddon and his staff at Queen's University Belfast for their help with synthesis of a number of ionic liquids and their insight into the project.

Alguacil F. J. and Lopez F. A. (1996) The extraction of mineral acids by the phosphine oxide Cyanex 923, *Hydrometallurgy*, **42(2)**, 245-255.

Baldwin W. H., Higgins C. E. and Soldano B. A. (1959) The distribution of monovalent electrolytes between water and tributyl phosphate, *Journal of Physical Chemistry*, **63(1)**, 118-123.

Berthod A., Ruiz-Angel M. J. and Huguet S. (2005) Nonmolecular solvents in separation methods: Dual nature of room temperature ionic liquids, *Analytical Chemistry*, **77**, 4071-4080.

Brennecke J. and Maginn E. (2001), Ionic liquids: innovative fluids for chemical processing, *AIChE Journal*, **47(11)**, 2384-2389.

Cytec data sheet, Cyphos® IL 104 phosphonium ionic liquid: Trihexyl(tetradecyl)phosphonium bis(2,4,4-trimethylpentyl)phosphinate.

De Beni G., Pierini G. and Spelta B. (1980) The reaction of sulphur dioxide with water and a halogen. The case of iodine: reaction in presence of organic solvents, *International Journal of Hydrogen Energy*, **5**, 141-149.

Dietz M. L. and Dzielawa J. A. (2001) Ion-exchange as a mode of cation transfer into room-temperature ionic liquids containing crown ethers: implications for the 'greenness' of ionic liquids as diluents in liquid-liquid extraction, *Chemical Communications*, **20**, 2124-2125.

Dietz M. L., Dzielawa J. A., Laszak I., Young B. A. and Jensen M. P. (2003) Influence of solvent structural variations on the mechanism of facilitated ion transfer into room-temperature ionic liquids, *Green Chemistry*, **5(6)**, 682-685.

Dietz M. L. and Stepinski D. C. (2005) A ternary mechanism for the facilitated transfer of metal ions into room-temperatures ionic liquids (RTILs): implications for the 'greenness' of RTILs as extraction solvents, *Green Chemistry*, **7**, 747-750.

Earle M. J., Esperanca M. S. S., Gilea M. A., Canongia Lopes J. N., Rebelo L. P. N., Magee J. W., Seddon K. R. and Widegren J. A. (2006) The distillation and volatility of ionic liquids, *Nature*, **439**, 831-834.

Ewan B. C. and Allen R. W. K. (2005) A figure of merit assessment of the routes to hydrogen, *International Journal of Hydrogen Energy*, **30**, 809-809.

Giaconia A., Caputo G., Ceroli A., Diamanti M., Barbarossa V., Tarquini P. and Sau S. (2007) Experimental study of two phase separation in the Bunsen section of the sulfur-iodine thermochemical cycle, *International Journal of Hydrogen Energy*, **32**, 531-536.

Holbrey J. and Seddon K. (1999), Ionic liquids, *Clean Products and Processes*, **1**, 223-236.

Huddleston J., Visser A., Reichert W., Willauer H., Broker G. and Rogers R. (2001) Characterisation and comparison of hydrophillic and hydrophobic room temperature ionic liquids incorporating the imidazolium cation, *Green Chemistry*, **3**, 156-164.

Ignat'ev N. V., Welz-Biermann U., Kucheryna A., Bissky G. and Willner H. (2005) New ionic liquids with tris(perfluoroalkyl)trifluorophosphate (FAP) anions, *Journal of Fluorine Chemistry*, **126**, 1150-1159.

Kertes A. S. (1960) Solute-solvent interaction in the system hydrochloric acid-water-tri-n-butyl phosphate, *Journal of Inorganic and Nuclear Chemistry*, **14**, 104-113.

Martak J. and Schlosser S. (2007) Extraction of lactic acid by phosphonim ionic liquds, *Separation and Purification Technology*, **57**, 483-494.

Mrnka M., Jedinakova V. and Sladkovska J. (1974) The extraction of HCl by TDPO in benzene, *Journal of Radioanalytical Chemistry*, **21**, 87-95.

Nomura M., Okuda H., Kasahara S. and Nakao S-I (2005) Optimisation of the process parameters of an electrochemical cell in the IS process, *Chemical Engineering Science*, **60**, 7160-7167.

Norman J., Mysels K., O'Keefe D., Stowell A. and Williamson D. (1978) Water splitting: The chemistry of the I_2-SO_2-H_2O reaction and the processing of the H_2SO_4 and HI products, *Miami international conference on alternative energy*, **8**, 3495-3526.

Sakurai M., Nakajima H., Amir R., Onuki K. and Shimizu S. (2000) Experimental study on side-reaction occurrence condition in the iodine-sulfur thermochemical hydrogen production process, *International Journal of Hydrogen Energy*, **25**, 613-619.

Stoyanov E. S. (1993) Molecular state of hydrochloric acid and its tribuyl-phosphate extracts, *Journal of Structural Chemistry*, **34(5)**, 714-721.

Widmer H. M. (1974) Solvation effects and ion association in solvent extraction systems. I. The thermodynamics of hydrochloric acid in the water-methyl isobutyl ketone system, *Journal of Physical Chemistry*, **74(17)**, 3251-3262.

Status of the I-NERI Sulfur-Iodine Integrated-Loop Experiment

By

Ben Russ, Wendy Sweet
General Atomics Corporation

R. Moore, G. Naranjo
Sandia National Laboratories

Max Helie, Nicolas Pons
French CEA

Sandia National Laboratories (SNL), General Atomics Corporation (GA) and the French Commissariat à l'Energie Atomique (CEA) are participants in a international collaborative effort to construct a laboratory-scale Sulfur-Iodine (S-I) process capable of producing 100-200 L/hr of hydrogen. This project is being conducted as an International Nuclear Energy Research Initiative (INERI) project supported by the CEA and the U.S. DOE Nuclear Hydrogen Initiative. The process has been constructed at General Atomics facility in San Diego, CA and testing is ongoing.

The S-I cycle consists of three main processing steps: thermal decomposition of sulfuric acid, production of hydrogen iodide and the decomposition of hydrogen iodide to produce hydrogen. SNL is responsible for the sulfuric acid decomposition section, the French CEA the hydrogen iodide production section and GA the hydrogen iodide decomposition section. The process sections are constructed on skids and are integrated through a chemical storage skid. All chemicals used in the process are transferred between the skids through the chemical storage skid with the exception sulfur dioxide that is transferred directly from the SNL skid to the CEA skid.

Phase I of this project involved integration of the three process sections into a complete S-I cycle. The initial integrated operation was performed in April of 2008 and is documented in the SAND Report titled "Results of Phase 1 Testing of the Sulfur-Iodine Integrated Laboratory-Scale Experiment". Phase II of the work includes completion of initial integrated shakedown operations, control verification tests (interface skid operations, production rate tests, shutdown tests), transient response tests (startup, shutdown, flow rate perturbations, mismatched flow rates), and cross talk tests (inter skid contamination and response) and process improvements or modifications as budgets permit. The results from phase I of the project along with additional stand-alone and integrated tests have identified issues for each process section that have been addressed in phase II work. These include redesign and replacement of major system components, corrosion control and modification of process monitoring. Several identified problems have led to very significant process modifications. However, for all three process sections, all major modifications have been completed.

The General Atomics hydrogen iodide section has undergone several modifications and has been successfully operated multiple times to produce hydrogen. The hydrogen iodide reactor has been replaced with a smaller unit to decrease the time for preheating the unit and reach equilibrium while operating. Using the smaller reactor, hydrogen has been produced at a rate of 10 to 75 L/hr. Problems with differential pressure cells used throughout the process for monitoring and control have been eliminated by replacing the units with higher quality units. Ta alloy corrosion problems were encountered due to hydrogen embrittlement of Ta vessels and piping. Most of these issues have been mitigated for the ILS with either modified operational conditions, or periodic replacement. Longer term solutions are being developed and will be tested in FY09.

The French CEA has completed major modifications to the Bunsen section and the section is ready for additionally testing of the Bunsen reactor at this time. Completed modifications include: installation of a new Bunsen reactor equipped with Raschig ring packing and differential pressure cells for enhanced process control and a redesign and modification of the system for transferring iodine from the chemical storage skid to the CEA skid. In addition to these modifications, additional integrated testing with the SNL acid decomposition section has been completed and the transfer of chemicals between the CEA section and the SNL sections is now considered routine. The CEA will begin testing the new Bunsen reactor within the next few weeks as part of integrated operation with the SNL process section.

In the SNL acid decomposition section, system modifications include incorporation of an acid neutralization step to the process cooling water system to prevent corrosion of the circulating water chiller used in the process, an increased efficiency of the acid concentrator and the addition of a pressure control system to maintain a constant pressure in the acid decomposer. The acid decomposition process has been successfully operated multiple times over the past five months in the stand-alone and integrated modes. In the last operations of the process, the effect of temperature on acid decomposition was determined at a pressure of 1 bar. The results indicate the production of sulfur dioxide begins to decrease between 750 and 800°C with a decrease in gas production of 50% at 700°C. No sulfur dioxide was detected at 650°C.

The GA hydrogen iodide decomposition section and SNL acid decomposition section of the S-I process are ready for additional integrated testing. All modifications have been completed and successful stand-alone tests have been performed with the new equipment. Major equipment modifications have been completed to the CEA section and the process is ready to begin testing of the newly installed Bunsen reactor. Testing of the Bunsen reactor will be performed in the stand-alone and integrated mode.

201c Initial Studies of the Feasibility of Performing the Bunsen Reaction In Ionic Liquids

Alex Yokochi and Nicholas AuYeung. Department of Chemical Engineering, Oregon State University, 207 Gleeson Hall, Corvallis, OR 97331-2702

The use of hydrogen as an energy carrier to be used for various purposes has been strongly considered, including for such applications as transportation fuels and residential/commercial/industrial combined heat-and-power. Of the many ways to produce hydrogen one can count processes such as water electrolysis and steam reformation of fossil fuels. One method that shows the promise of high effciciency coupled with the use of sustainable energy resources, such as nuclear or solar heat, is the Sulfur-Iodine thermochemical cycle, where water is thermolysed in a cascade of chemical reactions to produce hydrogen and oxygen.

A fundamental step of this chemical reaction cascade is a reaction where sulfur dioxide is oxidized by elemental iodine in the presence of water known as the Bunsen reaction, highlighted in yellow in Scheme 1 below.

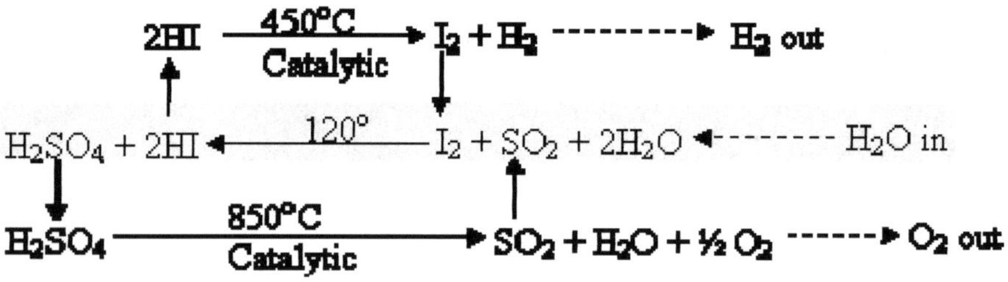

Scheme 1. Summary of Sulfur-Iodine thermochemical cycle with Bunsen reaction highlighted in yellow.

Currently, however, this reaction is carried out in a medium with a vast excess of water, due to the fact that the aqueous products, namely HI and SO_3 form a biphasic aqueous system, with the sulfur bearing stream spontaneously separating from the hydrogen iodide stream. While this simplifies this separation, it creates a need for the removal of the excess of water from the system, which becomes a critical problem. Several approaches to achieve this water removal have been proposed including reactive distillation and electrodialysis.

A cursory examination of the boiling points of the various reactants and products, listed in Table 1, suggests a different approach to the separation of the sulfur bearing stream from the iodide bearing one: taking advantage of the vapour pressure of the pure products.

Table 2. Boiling points of the chemical species involved in the Sulfur-Iodine cycle and its variants.

Species	Boiling point at 1 atmosphere
Water (H_2O)	100°C
Hydrogen Iodide (HI)	-35.4°C
Sulfur Dioxide (SO_2)	-10.1°C
Sulfur Trioxide (SO_3)	44.7°C
Sulfuric Acid (H_2SO_4)	337°C
Iodine	184.3°C
Ionic Liquids	Greater than 400°C

To enable this, the reaction must be carried out without the excess of water present, and in a solvent with negligible vapor pressure between the reaction temperature, occurring at close to room temperature, to the temperature at which the sulfur bearing species evolves from solution, which is up to 400°C. Our team has elected to investigate the feasibility of this reaction in an ionic liquid such as an alkyl-methyl-immidazolium salt of tetrafluoroborate or hexafluorophosphate.

These ionic liquids are miscible with water, excellent solvents, inert towards oxidation/reduction under the conditions described, and exist as stable liquids from well below -20°C to over 400°C.

The presentation will present initial feasibility results for this process.

201d Hydrogen Production by Sulfur-Iodine Thermochemical Cycle: Experimental Tests on Hydriodic Section Concerning Iodine Purification System

Raffaele Liberatore, Antonio Ceroli, Claudio Felici, Michela Lanchi, Salvatore Sau, Annarita Spadoni, and Pietro Tarquini. Ter Soltermsvil, ENEA, Via Anguillarese, 301, Rome, I-00123, Italy

The Sulfur-Iodine thermochemical cycle is one of the most promising routes currently under development for sustainable hydrogen production from water. It is based on three main reactions carried out at three different temperature levels (low, medium and high temperature):

- Bunsen reaction, exothermic at (80÷120) °C:

$2H_2O + I_2 + SO_2 \rightarrow H_2SO_4 + 2HI$

- Hydriodic acid decomposition, endothermic, at (300÷500) °C:

$2HI \rightarrow I_2 + H_2$

- Sulfuric acid decomposition, endothermic, at (800÷900) °C:

$H_2SO_4 \rightarrow H_2O + SO_3 \rightarrow H_2O + SO_2 + ½ O_2$

In the Bunsen section, water, iodine, and SO_2 at a temperature between 80 °C and 120 °C, produce two liquid immiscible phases: the superior one containing a mixture of H_2SO_4 and water, and the other including HI, I_2 and water.

In the Sulfuric acid decomposition section the Bunsen reaction superior phase, mainly composed of a H_2SO_4-H_2O mixture, is concentrated from (50÷57) %w/w to (90÷98) %w/w and decomposed in water, oxygen and SO_2 which is returned in the Bunsen section.

The Bunsen reaction inferior phase, instead, is sent to the hydriodic acid decomposition section where HI is decomposed in hydrogen and iodine. HI, residual H_2SO_4, and a very high content of iodine (~ 82 %w/w) compose this phase, since the Bunsen reaction is conducted with an excess of iodine to promote the separation of the two phases. A preventive SO_2 and iodine removal by an intermediate purification-separation step is necessary, in particular considering the very low decomposition reaction conversion rate (22 % at 480 °C) that requires a recirculation of these components.

Concerning hydriodic section, the most critical phase is represented by the separation-purification step of HI. Several techniques have been proposed on this purpose and their technical and economical sustainability have been demonstrated (Allen et al., Process Safety and Environmental Protection, 83 (B4) 343-350).

In the present work, in view of a forthcoming realization of an integrated loop plant at a laboratory scale, classic distillation process for I2 separation has been considered. This technique is a conventional way to separate HI and water from I2, in view of the high difference of the boiling temperature among the three elements (at room pressure: -35°C for HI, 100°C for H2O, 184 °C for I2).

In the Bunsen inferior phase, the ratio HI/(HI+H2O) generally approaches to 57 %w/w. For iodine free mixtures, this ratio corresponds to the azeotrope, characterized by a maximum temperature boiling point, corresponding to 127 °C (1 bar). Our tests and an Engels's work (H.Engels, K.F. Knoche, International Journal of Hydrogen Energy, 11, pp. 703-707,1986) showed that isotherms of different HI-H2O-I2 mixtures containing fixed I2 contents, have the minimum vapor pressure, defined as pseudo-azeotrope points, just near at this particular ratio between HI and HI+H2O.

In these conditions, it is plausible, by a distillation column, obtaining this value in the entire equipment, where the only iodine composition will vary from very low percentages, at the top, to 99 % w/w, at the bottom.

Nevertheless, opportunely dosing the reagents and in particular temperature conditions, it could be possible obtaining, by Bunsen reaction, HI-I2-H2O mixtures with HI content superior than the pseudo-azeotropic point, so that at the top of the column we could have only HI (although in low amount). On the contrary, if the amount of water loaded in the Bunsen reactor is too much, we could obtain HI-I2-H2O mixtures with a HI content inferior than the pseudo-azeotrope. In this case, the distillation column produces water at the top.

Since the purification-separation equipment must resist at a very aggressive environment and the column boiler need a great energy amount (see: Kasahara et al., "Journal of chemical engineering of Japan, vol.36, n°7, pp.887-889, 2003; Goldstein et al., "International Journal of Hydrogen Energy, 30 (2005) 619-626), this section affects for a high share the global process cost and efficiency. Therefore, a scrupulous design is suitable. In order to scale and optimize the distillation column, experimental HI-I2-H2O liquid-vapor equilibrium data for a wide concentration domain of the liquid phase, have been investigated.

These mixtures had I2 concentration in a range included from 0.2 %w/w to 90 %w/w and HI from 4 %w/w to 67 %w/w in the liquid phase.

The liquid phase composition versus temperature measurements demonstrated that temperatures show a different behaviour depending on the HI and I2 concentration.

If HI content is higher than pseudo-azeotrope: the I2 increase produces a temperature rising, on the contrary, the HI augment turns out a temperature lowering, because its concentration goes away from the binary azeotrope HI-H_2O.

HI augmentation, instead, in mixtures with HI minor than pseudo-azeotrope, rises the boiling temperature. Concerning iodine, experimental evidence showed a boiling temperature increase for mixtures with HI content near pseudo-azeotrope, but if this content is substantially lower (i.e.: 25 %w/w or 45 %w/w in iodine free basis) a light temperature reduction has been observed, at least for medium-low iodine contents.

In these last cases, some problems occur if we need to condense the vapor because it has a high content of water and small quantity of iodine (HI is very low). Since the iodine solubility in water is high only when in the mixture is being a remarkable HI content (C. F. Powell, I. E. Campbell, Journal of the American Chemical Society , Vol.: 69, pp.1227-1228, 1947), iodine precipitates and piles up on the condenser walls, causing its gradual obstruction.

In every case, iodine high content mixture samplings need an insulated and thermostatic sample system in order to avoid the sample cooling and the consequent iodine solidification.

At the end of the liquid-vapor equilibrium characterization (at room pressure), the present study has been focused on distillation preliminary tests of the mixtures coming out from Bunsen reaction inferior phase. These tests have been executed, in batch, by two kind of Pyrex equipment: one Vigreux column (Height = 600 mm, diameter = 32 mm) and one packed column (Height = 600 mm, diameter = 32mm) with Rasching rings (diameter = 7 mm). Other tests, using, as feed, inferior solutions, directly coming from Bunsen reactor, have been performed in continuous by a packed column (with Rasching rings) bigger than the others (Height = 1000 mm, diameter = 50 mm) and equipped by pumps and an external condenser.

The condenser showed obstruction problems, concerning mixtures with HI content substantially lower than the pseudo-azeotrope, due to the cited precipitation of the iodine. In these last cases, indeed, at the top of the column, water essentially went out, HI being absent. The Bunsen inferior phase, generally, includes small amounts of H_2SO_4 (Sakurai et al., Int. Journal of Hydrogen Energy, 25 (2000) 605-611) so, if they entered the column, go out as distillate together water, before the HI-H_2O mixture. Therefore, it is evident the need to control the feed composition be appropriate and to insert a purification equipment in order to eliminate the H_2SO_4 and to avoid the water excess.

However, these tests confirmed the possibility to separate iodine from Bunsen inferior phase and the results are showed in this work.

201e Process Simulations of Reactive Distillation for Hydrogen Iodide Decomposition In the Sulfur-Iodine Cycle

John P. O'Connell and James E. Murphy. Chemical Engineering, University of Virginia, 102 Engineers' Way, Charlottesville, VA 22904

Contemporary scenarios for future energy supply include using enormous amounts of hydrogen as an energy carrier for fuel cells and as a reagent for upgrading heavy oils and converting coal to liquid fuels. The Sulfur-Iodine cycle is a closed-cycle, all fluid-phase process to thermochemically decompose water into hydrogen and oxygen that may be the most energy efficient among the many proposed. Current process simulation studies suggest that the energy requirements are unacceptably high, especially in the section where H_2 is formed from HI and separated from the recycled I_2 by reactive distillation. The reliability of these analyses, and the testing of creative alternatives, is hampered by the lack of reliable data to build the complicated properties models for these systems.

To make an informed decision with regard to the ultimate feasibility of the S-I process, accurate predictions of phase and reaction equilibria are essential to yield reliable process simulations for equipment and energy requirements. A new properties model, described elsewhere, has been utilized to find optimal configurations and conditions for this section of the S-I process. Significantly different results are found compared to the literature; these suggest that alternative arrangements could lead to improved efficiencies for the process. This paper will describe our methodology and current findings about the RD column, as well as indicate potential opportunities in processing for improved efficiency.

201f Simulation of Flowsheet Options for the HI Concentration and Decomposition Section of the Sulfur-Iodine Thermochemical Cycle

George M. Bollas[1], Patrick Smadbeck[2], and Paul I. Barton[2]. (1) Department of Chemical Engineering - Process Systems Engineering Laboratory, Massachusetts Institute of Technology, 77 Massachusetts Ave., RM 66-363, Cambridge, MA 02139, (2) Chemical Engineering, Massachusetts Institute of Technology, 77 Massachusetts Ave., RM 66-363, Cambridge, MA 02139

Thermochemical water splitting is being developed as a potentially energy-efficient hydrogen production process. Of the proposed large-scale hydrogen production processes, the Sulfur-Iodine (SI) thermochemial cycle shows potential to be one of the more cost-effective. In this thermochemical cycle, iodine and sulfur dioxide react with water to form hydrogen iodide and sulfuric acid. The resulting mixture readily splits into two immiscible liquid phases, allowing for effective separation of the acids. Sulfuric acid is then decomposed at high temperatures (~850°C) releasing sulfur dioxide, which is recycled, and oxygen. The hydrogen iodide is similarly decomposed at about 350°C to produce hydrogen and iodine, which is recycled.

The main purpose of this work is to study ways to optimize the third section in this cycle, the concentration and decomposition of hydrogen iodide. This section limits the overall efficiency of the SI thermochemical cycle due to certain conditions present during operation. First, the H_2O-HI system exhibits an azeotrope, which is a bottleneck for their separation. Second, the H_2O-HI-I_2 system has two separate regions of liquid immiscibility, which makes the thermodynamic modeling of this system an intriguing task. Third, the H_2O-I_2 system has a strong immiscibility in the liquid phase and the mechanism of the HI decomposition reaction in the vapor phase is uncertain.

De-bottlenecking of this section is suggested to be achieved through three possible separation options. The first option is using extractive distillation [1] which introduces phosphoric acid to the system to induce the separation of iodine allowing for simple distillation of HI. HI is then decomposed at about 450°C to produce H_2. Separation is then facilitated by membranes to produce pure hydrogen gas. There are distinct advantages to this system, as it simplifies the model being considered, but the disadvantages are severe. By adding phosphoric acid the efficiency of the system drops significantly making the process costly and unattractive. Moreover, the thermodynamic modeling of the multi-electrolyte HI, H_3PO_4 solution is quite complicated. The second separation option being considered is electrodialysis [2,3] which should be able to concentrate the HI_x mixture beyond its azeotropic point more easily allowing for the decomposition and extraction steps. Excess HI is removed by distillation, and the steps for decomposition and extraction are the same as in extractive distillation. This process has several advantages, as it allows for an increase in efficiency of the above process, but the disadvantages are similar to the extractive distillation process. The third separation option being examined is reactive distillation [4], where both HI_x distillation and HI decomposition occur simultaneously at 350°C. A liquid-gas equilibrium occurs in the middle of the column, I_2 is solubilized in the liquid phase, and H_2 is recovered from the top of the column. This process has a distinct disadvantage in that the column has a very large feed stream and, therefore, very high heat demand. Moreover, it requires a complex chemistry-thermodynamic model, a problem augmented by the limited experimental data for the HI_x mixture. Its advantages are that it would provide a cost-effective, efficient method to perform the limiting process in this cycle.

A great deal of problems in the study of the advantages and disadvantages of each process have to do with the high degree of uncertainty regarding the thermodynamics of the HI-H_2O-I_2 mixture and the kinetics of the HI decomposition. A detailed, accurate and consistent thermodynamic model, verified with the majority of experimental data available [5], is required for a reliable study. In the past, problems have been identified with the thermodynamic models being used for the extreme conditions of the SI thermochemical cycle [6]. Using the concepts of hydration and polyiodide ions formation a new thermodynamic model has been developed to address the specific needs of this section. The model was implemented in the aspenONE simulation software and detailed flowsheets for the HI section were examined. Efficiency and feasibility studies were performed on the reactive distillation option, considered the most promising of the separation processes examined.

References

1. Norman JH, Besenbruch GE, Brown LC, O'Keefe DR, Allen CL. Thermochemical water-splitting cycle. Bench-Scale Investigations and Process Engineering. 1982;Final Report - General Atomics.

2. Arifal, Hwang G-J, Onuki K. Electro-electrodialysis of hydriodic acid using the cation exchange membrane cross-linked by accelerated electron radiation. Journal of Membrane Science. 2002;210:39-44.

3. Hwang G-J, Onuki K. Simulation study on the catalytic decomposition of hydrogen iodide in a membrane reactor with a silica membrane for the thermochemical water-splitting IS process. Journal of Membrane Science. 2001;194:207-215.

4. Roth M, Knoche KF. Thermochemical water splitting through direct Hi-decomposition from H_2O/HI/I_2 solutions. International Journal of Hydrogen Energy. 1989;14:545-549.

5. Doizi D, Dauvois V, Roujou JL, Delanne V, Fauvet P, Larousse B, Hercher O, Carles P, Moulin C, Hartmann JM. Total and partial pressure measurements for the sulphur-iodine thermochemical cycle. International Journal of Hydrogen Energy. 2007;32:1183-1191.

6. Bollas GM, Kazimi MS, Barton PI. Detailed Modeling Of The Thermodynamics Of The Sulfur-Iodine Thermochemical Cycle. AIChE Annual Meeting.2007; Salt Lake City, Utah, USA.

273a A Membrane Process for the Concentration of Sulfuric Acid for the Sulfur-Based Thermochemical Cycles

Frederick F. Stewart and Christopher J. Orme. Chemical Sciences, Idaho National Laboratory, P.O. Box 1625, Idaho Falls, ID 83415-2208

In the sulfur-iodine (S-I) thermochemical cycle for hydrogen production, the chemistry is dictated by the oxidation and reduction behaviors of sulfur and iodine. Hydrogen is produced from the decomposition of hydriodic acid (HI). To form HI, the Bunsen reaction is conducted that produces an HI and sulfuric acid mixture, which is then separated. Inputs into the Bunsen reaction are water, SO2 and I2. Thus, sulfur is cycled between its +4 and +6 oxidation states, SO2 and H2SO4, respectively.

Necessary SO2 for the Bunsen reaction is provided by the decomposition of sulfuric acid. Typically, the decomposition of sulfuric acid is performed over a catalyst, such as platinum. However, even at the highest temperatures (~850 °C) and using the most active catalyst materials, the reaction remains equilibrium limited, which limits the overall conversion and creates significant amounts of unreacted sulfuric acid that requires recycle. From the sulfuric acid that is decomposed during the reaction, water is liberated, which occurs when the acid initially dissociates to SO3 and water on heating. Thus, the effluent from the decomposer is the desired SO2, and diluted sulfuric acid. To recycle the acid, water must be removed. Distillation is a natural option and can perform the concentration if enough heat is applied. An issue with distillation is thermally induced corrosion of equipment, which surprisingly is a more significant issue for dilute sulfuric acid than for concentrated acid, especially at the vapor-liquid interface. Thus, an alternate technology was sought that could perform the separation at lower temperature.

Previous work applied to S-I cycle was focused on HI concentration. Nafion-® membranes were found to be effective in the removal of water from concentrated HI and HI/I2 solutions using pervaporation at temperatures as high as 140 °C. Performance of the membranes was considered excellent with both high flux rates and separation factors with no significant degradation of the membranes. This technology was applied to sulfuric acid with similar results. Flux of water through the membranes was found to be commercially competitive with high selectivity. In this paper, performance of two differing thicknesses of Nafion® will be discussed including a stability study of the membranes using Dynamic Mechanical Analysis. Our results indicate that the Nafion® materials are stable and perform predictably; however there is some slight embrittlement of the membranes upon exposure to acid. The embrittlement has been characterized instrumentally, but does not appear to have an effect on water transport through the membranes as reflected by permeability and selectivity data. Thus, this technology may be applicable to the S-I cycle and to any other cycle employing sulfur. Moreover, there also may be industrial applications considering the ubiquitous nature of sulfuric acid as a commodity.

273b Application of Nafion Membranes to the Sulphur Iodine Cycle

Rachael H. Elder, Najwa M. Elbakhbakhi, Geofrey H. Priestman, and Ray W. K. Allen. Department of Chemical and Process Engineering, University of Sheffield, Mappin St, Sheffield, S1 3JD, United Kingdom

Across the World there is a significant R&D effort concentrated on the Sulphur Iodine cycle, with a view to efficient hydrogen production from water. This paper investigates the application of Nafion® membranes to both the HI_x and H_2SO_4 processing sections.

Flux and selectivity data using a stirred batch process for the dewatering of $HI\text{-}H_2O\text{-}I_2$ solutions have previously been presented. Here, this work is continued by using a continuous flow set up similar to those used in industrial membrane processes. Excellent separation factors were achieved and the data collected correlated closely with that from previous experiments.

Simulations have been performed to identify the potential benefits of dewatering the H_2SO_4 solution leaving the Bunsen section. An experimental programme has also been carried out, assessing the stability and applicability of Nafion® membranes for water separation by pervaporation. Different thickness Nafion® membranes have been investigated, across a range of temperatures and acid concentrations, and the results are presented here.

Application of Radiation-Grafted Polymer Electrolyte Membranes to the Bunsen Reaction

Nobuyuki Tanaka, Masaharu Asano, Kaoru Onuki and Yasunari Maekawa,
Japan Atomic Energy Agency

Abstract

Bunsen reaction was performed electrochemically using cation exchange membranes prepared by a radiation-induced grafting method for application to the iodine-sulfur thermochemical water-splitting cycle. The reaction experiments using sulfuric acid dissolving sulfur dioxide as the anolyte and hydriodic acid dissolving iodine as the catholyte verified the progress of the aimed reaction. Compared with Nafion 117, the radiation-grafted membranes showed lower electric resistance in the reaction condition, whereas they were less selective to proton permeation.

Introduction

Thermochemical water-splitting methods have been attracting a lot of interests because of its potential of efficient and economic hydrogen production, coupled with High Temperature Gas-Cooled Reactors (HTGR). Among many cycles proposed so far, those using iodine- and sulfur-compounds as process chemicals have been studied most intensively, the basic chemical reactions of which are as follows;

$$I_2 + SO_2 + 2H_2O = 2HI + H_2SO_4 \quad \text{[Bunsen reaction]} \quad [1]$$
$$2HI = H_2 + I_2 \quad [2]$$
$$H_2SO_4 = H_2O + SO_2 + 1/2\ O_2 \quad [3]$$

The reaction [1], known as Bunsen reaction, proceeds exothermically as SO_2 gas absorption reaction producing HI and H_2SO_4, and, then, decomposed endothermically at elevated temperatures producing H_2 and O_2, respectively. I_2 and SO_2 produced in the reactions [2] and [3], respectively, are reused in the Bunsen reaction. As the net result, H_2O is decomposed into H_2 and O_2. Here, the reaction [3] requires the highest temperature of over 800°C, which is well matched with the temperature range of HTGR.

Most of the present studies on the water-splitting process consider utilization of the liquid-liquid phase separation (LL separation) phenomenon for separating HI and H_2SO_4 produced by the Bunsen reaction, which was found by researchers of General Atomics (1). The LL separation occurs in the presence of excess amount of iodine and produces a heavier liquid phase containing mainly HI, I_2, H_2O and a lighter liquid phase containing mainly H_2SO_4 and H_2O. This method offers an easy and simple way of the separation. However, it requires large inventory and large circulation of iodine.

The present study concerns the Bunsen reaction using an electrochemical cell equipped with a cation exchange membrane. Fig. 1 shows the concept of the reaction system. The electrode reactions are as follows

Anode : $SO_2 + 2H_2O = H_2SO_4 + 2H^+ + 2e^-$ [4]

Cathode: $I_2 + 2H^+ + 2e^- = 2HI$ [5]

Total: $I_2 + SO_2 + 2H_2O = 2HI + H_2SO_4$ [Bunsen reaction] [1]

By using the separation membrane, an excess amount of iodine is not required for the separation. Dokiya et al. (2) first proposed the idea of using electrochemical cell and demonstrated the aimed separation. Recently, Nomura et al. (3) verified the production of sulfuric acid and hydriodic acid, whose concentrations were higher than those attainable by the spontaneous thermal reaction. However, in their experiments using Nafion, very high cell voltage was necessary for operating with the concentrated solutions, and sulfur formation was observed. In this paper, applicability of polymer electrolyte membranes prepared by a radiation-induced grafting method (4) was examined preliminarily.

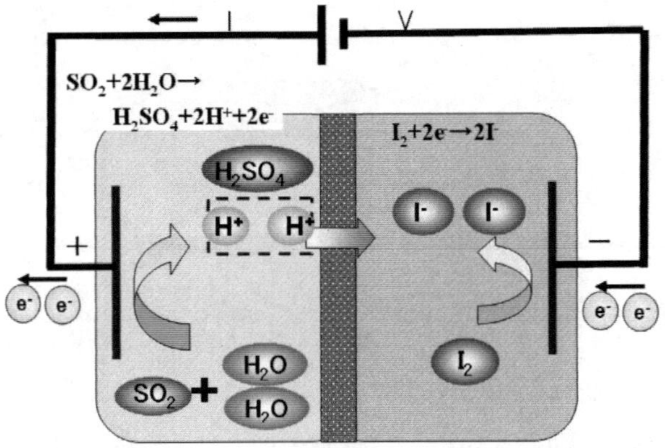

Fig. 1 Concept of electrochemical Bunsen reaction electrochemical cell

Experimental

Chemicals and membranes

Hydriodic acid (55-58wt%, Kanto Chemical Co., Inc.), iodine (>99.8wt%, Kanto Chemical Co., Inc.) and Sulfuric acid (97wt%, Wako Pure Chemical Industries, Ltd. and 47wt%, Kanto Chemical Co., Inc.) were used as received for prepared test solutions. SO_2 gas (>99.9%, Sumitomo Seika Chemicals Co., Ltd.) was used to prepare the anolyte solution by bubbling into sulfuric acid for one hour.

In the preparation of grafted membranes, poly-(ethylene-co-tetrafluoroethylene) (ETFE) film (thickness of 50 μm, kindly supplied from Asahi Glass Co.) was used as a base film for grafting. ,*P*-methlstyrene (MeSt, 99%, Sigma-Aldrich Japan K.K.) and *p-tert*-buthylstyrene (tBuSt,

93%, Sigma-Aldrich Japan K.K.) were used for grafting monomers. The sulfonic acid was introduced by the sulfonation reaction of the grafts in the membranes. Details of the membrane preparation and characterization were described elsewhere (4). Table 1 lists the properties of membranes used in this study. Nafion 117 (E.I.du Pont) was used as the reference membrane.

Table 1 Properties of the membranes prepared

ID	Monomer	IEC [mmol/g]	Conductivity [mS/cm]	Water uptake [wt%]
RG1	MeSt	1.4	71	39
RG2	MeSt/tBuSt(1:1)	1.4	82	44
RG3	tBuSt	1.0	22	22

0.1 mol/L sodium thiosulfate solution, 0.1mol/L sodium hydroxide, 0.5mol/L Iodine solution were purchased from Kanto Chemical Co., Inc. and used as received for the concentration analysis of test solutions.

Electrochemical Bunsen reaction

The experimental setup consisted of an electrochemical cell, two glass reservoirs for anolyte and catholyte solutions, two peristaltic pumps (RP-1000, Tokyo Rikakikai Co. Ltd.), and a DC power supply (HA-305, Hokuto Denko Corp.). The cell was a filter-press type one (FC-05-02, ElectroChem, Inc.) with two rubber gaskets made of Kalrez® (thickness of 0.5mm). Carbon felts (Toyobo Co. Ltd.) were used as the electrode by inserting in between membrane and carbon current collectors. The effective membrane area was 5cm^2. Rubber heaters were used to control the operation temperature. Glass reservoirs were connected with each compartment of the cell using Teflon® and Fluran® tubes.

In the experiment, 300g of a HIx solution with concentrations of [HI]=[I_2]=10 mol/kg was fed to the catholyte reservoir, and 300g of SO_2 gas saturated sulfuric acid with H_2SO_4 molality of 10 mol/kg was fed to the anolyte reservoir. The solutions were heated to 40°C while circulating to the cell for about one hour. Then, direct current was supplied to the cell in a galvanostatic mode with current density of 200 mA/cm^2 for 3 hours. The catholyte and the anolyte solutions were sampled every 1 hour and the concentrations were measured by alkalimetry and iodometry.

Result and Discussion

Fig. 2 shows the changes of HI and I_2 molality in catholyte and H_2SO_4 molality in anolyte observed in the experiment using RG1 membrane. The change of SO_2 concentration is also shown in the same figure. As seen in the figure, in the course of operation, the HI molality in catholyte

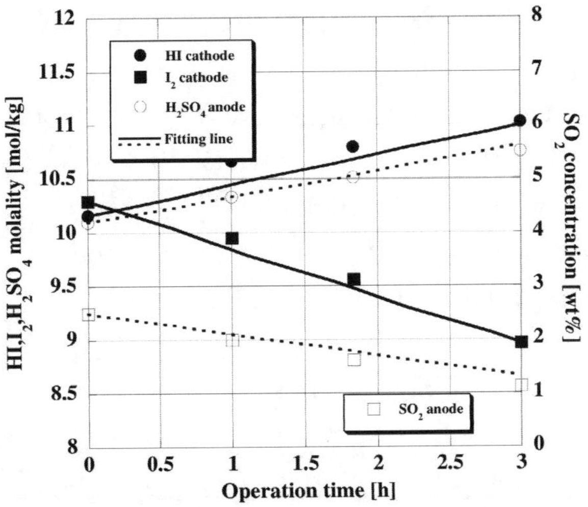

Fig.2 Concentration change in the electrochemical Bunsen reaction using RG1 membrane

Anolyte: SO_2 saturated sulfuric acid, $[H_2SO_4]$=10mol/kg

Catholyte: HIx solution, [HI]=$[I_2]$=10mol/kg

Temperature: 40°C

Current Density: 200 mA/cm^2

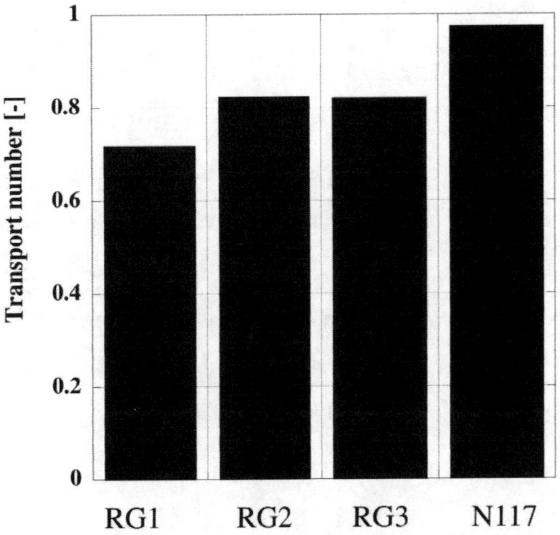

Fig. 3 Observed transport number

Experimental conditions are the same as in Fig.2

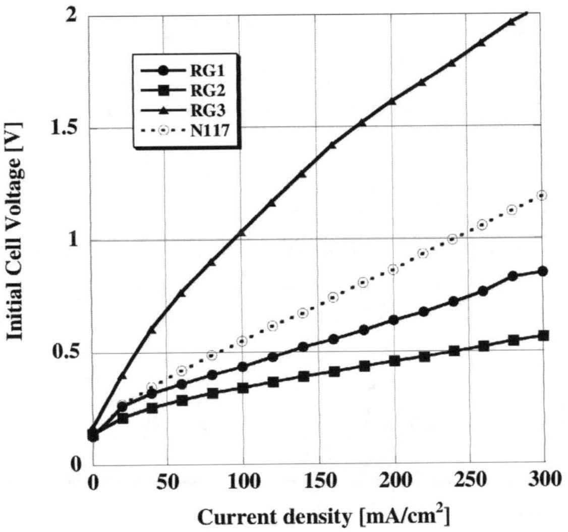

Fig. 4 Cell voltage vs current density

Experimental conditions are the same as in Fig.2

increased from 10.1 to 11.0mol/kg and the H_2SO_4 molality in anolyte increased from 10.1 to 10.7mol/kg, whereas the I_2 and the SO_2 concentrations decreased. These changes are qualitatively explained by assuming the reactions [4] and [5], indicating the progress of expected Bunsen reaction. At the same time, in the course of operation, the anolyte solution became yellow and the membrane surface faced to anolyte solution became white. It was presumed that precipitation of sulfur caused these phenomena.

In order to discuss the sulfur formation mechanism, transport number of proton (t_+) was estimated. It was defined as follows,

$$t_+ = \frac{F\Delta M_{HI}}{I \cdot t}$$

where, F, ΔM_{HI}, I, t denote Faraday constant, change of HI quantity in catholyte [mol], electric current supplied [A], operation time [s], respectively. The estimation of transport number was carried out by simulating the observed change of HI molality.

Fig. 3 shows the obtained transport numbers. The fitting line used this parameter was showed in Fig. 2. Nafion117 showed high value of 0.98, which was in accord with the reported one (3), whereas the radiation-grafted membranes showed lower value of 0.7 - 0.8. Also, it was observed that the lower the transport number was, the clearer the color changes were in anolyte and in membrane. These results suggest that iodide ion (I^-) permeated from catholyte to anolyte reacted with SO_2 to produce sulfur. The reaction equation might be "$SO_2 + 4HI = S + H_2O + 2I_2$" (3,5).

Fig. 4 shows the relationships between cell voltage and current density. The cell voltage was largest in the experiment using RG3 membrane, followed by RG1 and RG2 membranes. The cell voltages in the latter two cases were lower than that in the case of Nafion117.

It is interesting to note that RG2 membrane showed highest transport number and lowest electric resistance, as well, among the membranes prepared. Especially, the observed cell voltage was about half of that of Nafion. RG2 was prepared by graft co-polymerization of *p*-methylstylene and *t*-butylstylene, and exhibited the similar low electric resistivity as RG1 that featured *p*-methylstylene and the similar high ion selectivity as RG3 featuring *t*-butylstylene. The results suggest the possibility of further improvement of the membrane performance by optimizing membrane structure.

Conclusion

The radiation-grafted membranes were employed for the Bunsen reaction using an electrochemical cell equipped with carbon electrodes. The HIx solution with concentrations of $[HI]=[I_2]=10$ mol/kg in catholyte and SO_2 gas saturated sulfuric acid with H_2SO_4 molality of 10 mol/kg in anolyte were employed. The reaction was examined with current density of 200mA/cm^2 at 40°C. The results are summarized as follows.

1. The proton transport numbers of the radiation-grafted membranes (0.7-0.8) were lower than that of Nafion117 (0.98). Indication of sulfur formation due to the low ion selectivity was observed in anolyte.
2. Some radiation-grafted membranes showed lower electric resistance than Nafion117. The observed lowest cell voltage was about half of that of Nafion117.
3. A radiation-grafted membrane prepared by graft co-polymerizing *p*-methylstylene and *t*-butylstylene into ETFE film showed highest ion selectivity and lowest electric resistance among the membranes tested.

Improvement of the membrane performance is expected by optimizing the grafting conditions such as a base polymer and monomer structures, and a degree of grafting.

Acknowledgment

The authors express their gratitude to Mr. K. Ikenoya for the implementation of experiments. This work was supported by a Grant-in-Aid for Scientific Research from the Ministry of Education,

Culture, Sports, Science and Technology of Japan.

References

1. J.L. Russell et al., (1976) "Water splitting – A progress report," *Proc. 1st World Hydrogen Energy Conf.*, Miami-Beach, Florida, USA, March 1-3, 1976, 1A105-1A124.
2. M. Dokiya et al., (1979) "Thermochemical Hydrogen Production – Part V. A feasibility study of the Sulfur Iodine Cycle," Int. J. Hydrogen Energy, 4, pp.267-277
3. M. Nomura et al., (2004), "Development of an Electrochemical Cell for Efficient Hydrogen Production through the IS process," *AIChE Journal*, 50, pp.1991- 1998.
4. J. Chen et al., (2006) "Preparation and characterization of chemically stable polymer electrolyte membranes by radiation-induced graft copolymerization of four monomers into ETFE films," *J. Membr. Sci.*, 269, pp.194-204.
5. M. Sakurai et al., (2000) "Experimental study on side-reaction occurrence condition in the iodine-sulfur thermochemical hydrogen production process," *Int. J. Hydrogen Energy*, 25, pp.613-619.

Potential for using High Temperature Membranes in the Sulphur Trioxide Decomposition Stage of a Thermochemical Cycle

Ian Atkin[1], Geof H. Priestman[1], Derek C. Sinclair[2] and Ray W. K. Allen[1]
[1]Department of Chemical and Process Engineering, University of Sheffield, Mappin Street, Sheffield, S1 3JD, United Kingdom
[2]Department of Engineering Materials, University of Sheffield, Mappin Street, Sheffield, S1 3JD, United Kingdom

Introduction

The sulphur-iodine (SI) and Hybrid Sulfur (HyS) cycles are promising processes for massive scale hydrogen production. These thermochemical cycles take water as a feedstock and produce only hydrogen and oxygen, with no additional waste streams. Both cycles employ the high temperature thermal decomposition of sulphuric acid.

The sulphuric acid thermal decomposition process has two reaction steps. The sulphuric acid is first decomposed to sulphur trioxide and water. Equilibrium composition calculations for this first step show that very high yields of SO_3 are achieved at 600°C and 1 bar from a pure H_2SO_4 feed. Sulphur trioxide then further decomposes to sulphur dioxide and oxygen. The equilibrium position for SO_3 decomposition is such that temperatures higher than 600°C are required for high yields. From a pure H_2SO_4 feed at 1 bar, the SO_2 yield is around 52% at 700°C and 87% at 900°C.

The use of a membrane separation in the SO_3 decomposition process that could enhance the yield of SO_2 at a given temperature is examined in this work. This would allow the operation of such a process with a lower peak temperature, enabling the use of more readily available heat sources. Alternatively, it could allow greater reactor throughput for a given heat duty.

Continuous removal of a proportion of the O_2 produced by SO_3 decomposition would cause the SO_3 decomposition equilibrium position to be shifted such that a greater proportion of the SO_3 would be decomposed than would be the case for a simple equilibrium process.

There are two types of membrane that could be suitable, dense and microporous. Dense oxygen transport membranes transport oxygen through their crystal lattice by ionic conduction. Microporous membranes, where the mean pore diameter is much smaller than the mean free path of the gas molecules, exhibit Knudsen regime flow, whereby the separation factors between components are equal to the ratio of the square roots of the molecular weights of each component.

Notional layouts used for process simulations

Process simulations were carried out using two notional layouts: a simple catalytic reactor, with a membrane separator downstream and retentate recycling; and use of a notional catalytic membrane reactor (CMR) in lieu of the simple catalytic reactor.

A downstream separation membrane would permit a stream of non-equilibrium composition to be removed from the process. A fraction of the retentate is then recycled to the reactor feed, with the remaining retentate removed as the process product. This process is illustrated in Figure 1.

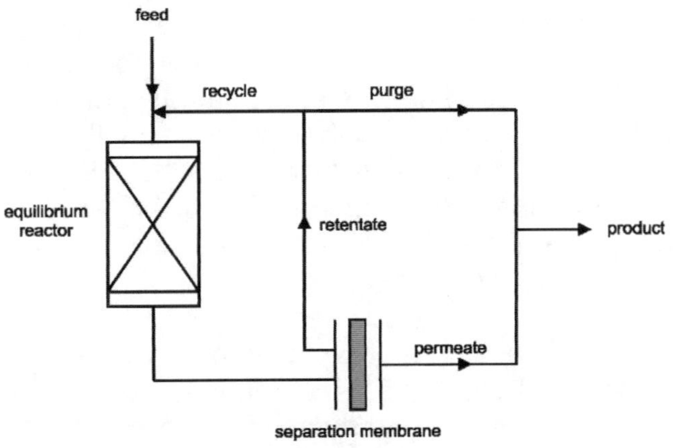

Figure 1: Schematic of downstream membrane separation/recycle process

The notional catalytic membrane reactor, as depicted in Figure 2, employs a number of tubular membranes inserted into a catalyst bed. The membranes selectively allow some products to permeate out of the catalyst bed, continuously shifting the reaction equilibrium and therefore potentially allowing a much greater yield of SO_2 than would be possible using a single equilibrium reactor.

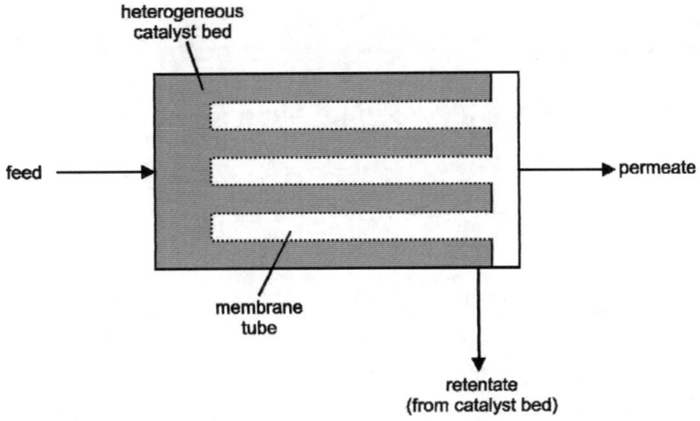

Figure 2: Schematic of catalytic membrane reactor (CMR) process

Forsberg [1] carried out a number of process simulations on the application of CMRs to the sulphuric acid thermal decomposition process. The simulations were made on the basis of a CMR with six theoretical 'stages', each allowing the process gas to come to equilibrium before performing a theoretically perfect separation. The simulations were replicated in order to validate the process simulation package used here, *ProSimPlus*.

Forsberg's work has been extended by carrying out a number of process simulations using both the CMR and downstream separation processes. Dense oxygen separation membranes and Knudsen regime porous membranes have been simulated with each process.

Specification for simulations

Each of the proposed process configurations have been modelled using the *ProSimPlus* steady state process simulation package. The components used for the simulations were H_2SO_4, H_2O, SO_2, SO_3 and O_2. A stream consisting of 92.7wt% H_2SO_4 and 7.3wt% H_2O entered the process and was subsequently heated to the required temperature. This acid concentration mirrors that specified by the original General Atomic flowsheet for the SI cycle acid decomposition section feed [2]. Simulations were carried out at 900K $\geq T \leq$ 1300K in 100K steps and 1 atm (1.01325 bar(a)) $\geq P \leq$ 30 bar(a). The reactors are modelled as gas phase equilibrium reactors performing a Gibbs free energy minimisation at the specified process temperature.

The dense oxygen separation membranes were simulated for the separation ratio, having values, F = 10, 100, 1000 and 10000. F is defined as:

$$F = \frac{\left(\frac{[O_2]}{[x]}\right)_{\text{permeate}}}{\left(\frac{[O_2]}{[x]}\right)_{\text{feed}}}$$

where x is each of the other species present. For dense membranes, these separation factors basically represent cross-membrane leakage, either as a result of membrane defects or poor sealing. The porous membranes were simulated as exhibiting Knudsen regime separation factors.

The overall yield of SO_2 was recorded for each simulation. For these simulations the SO_2 yield is equal to:

$$Y = \frac{\text{mols of } SO_2 \text{ in product}}{\text{mols of } H_2SO_4 \text{ in feed}}$$

Catalytic membrane reactor simulations

These simulations extended Forsberg's CMR simulation work by examining the SO_2 yield that could be achieved using CMRs of two to six theoretical equilibrium/separation stages. Additionally, separations with finite selectivity were employed. Figure 3 shows a typical *ProSimPlus* process layout for six theoretical 'stages'. It can be seen that the feed is heated then fed into a train of equilibrium reactors that are each followed by a separator, operating to a given selectivity. The products from each stage are mixed to produce the final product.

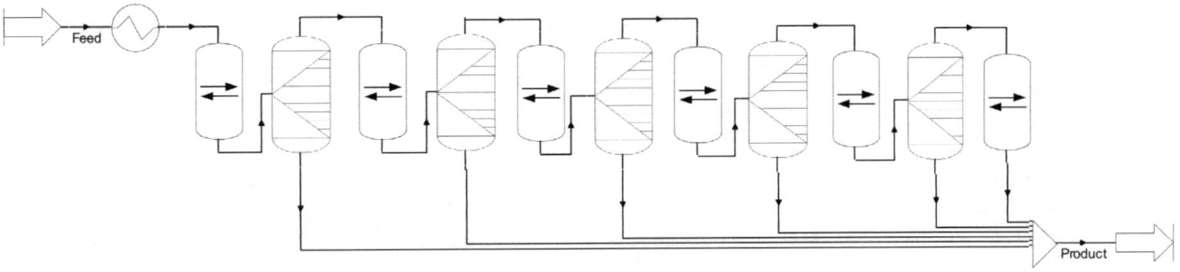

Figure 3: Typical *ProSimPlus* process layout for catalytic membrane reactor process simulations showing six theoretical equilibrium stages

As shown in Figure 4, increasing the number of theoretical stages beyond six produced little increase in SO_2 yield with the dense membrane simulations, whilst using more than two theoretical stages with microporous selectivity gave little improvement. This can be attributed to the flow rates through each stage. With the microporous membranes, the selectivity was sufficiently low that only a small proportion of the process flow was retained after the first two separation steps – most of the flow had "permeated" the membranes and passed out of the process. Because of this, the subsequent stages had very low flow rates and hence had little effect on the process as a whole. The selectivities relating to dense membranes were higher and hence a greater flow reached the later stages of the "train". Even here though, the flow rate was small once the eighth theoretical stage was reached.

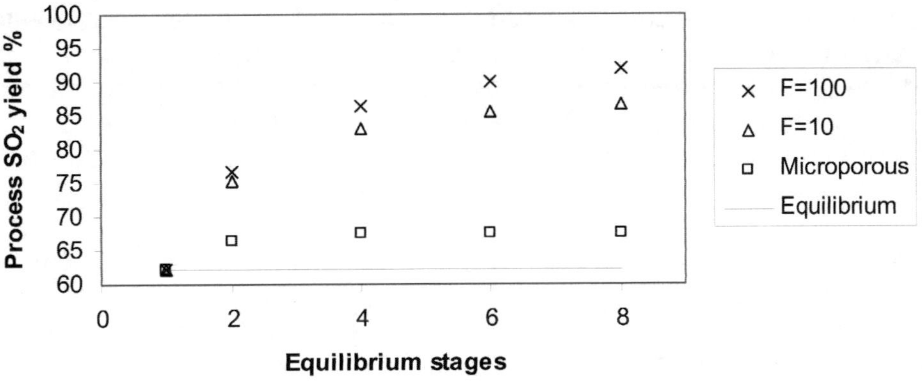

Figure 4: *ProSimPlus* CMR process simulations at 1 atm, 1000K and 92.7wt% H_2SO_4 feed

As the process temperature increases, the difference between the equilibrium SO_2 yield and the improved yields calculated by these simulations decreases (see Figure 5.) Because the equilibrium SO_2 yield increases with temperature, the scope for yield improvements decreases. The simulations suggest that at 1 atm and 1000K, using dense oxygen separation membranes and six theoretical stages, SO_2 yield could be increased from 62% to 85% or more. However, the same system at 1300K has an equilibrium yield of 95% and a dense membrane CMR yield of 99%, a higher absolute yield but a smaller improvement. It is because of this that it is thought that the CMR would be most beneficial to processes operating the sulphuric acid decomposition step at relatively low temperatures.

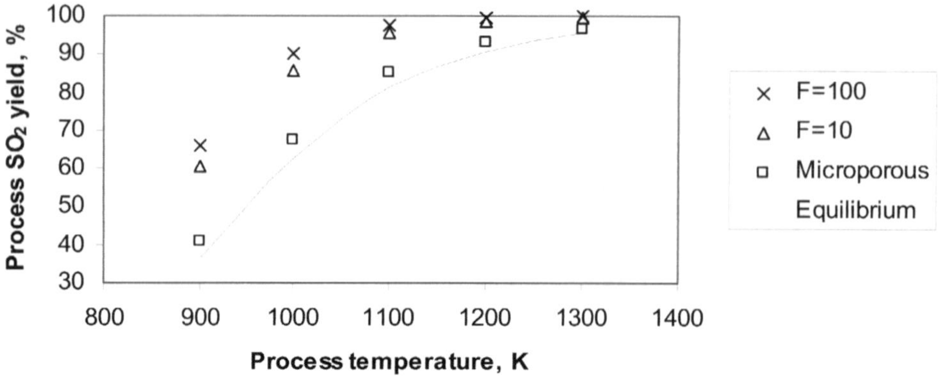

Figure 5: *ProSimPlus* CMR process simulations with six theoretical equilibrium/separation "stages", 1 atm and 92.7wt% H_2SO_4 feed

Downstream separation/recycle process simulations

The second part of the simulation work examined the potential for use of a downstream membrane separation and subsequent recycle, as opposed to the more complex CMR design.

The *ProSimPlus* simulations replicate the process depicted in Figure 1. The initial heated feed passes into the equilibrium reactor. A proportion of the reactor outlet stream passes through the membrane. A fraction of the membrane retentate is recycled and the remainder of the retentate is mixed with the membrane permeate to produce the process outlet stream. Recycle fractions of 20% $\geq R \leq$ 80% in 20% steps were used. It was assumed that 75% of the oxygen entering the separation step permeated through the membrane.

It was calculated that SO_2 yield increases with increasing process temperature, increasing recycle ratio and increasing membrane separation factor. As with the CMR simulations, the SO_2 yield increase over the equilibrium value is greatest at lower temperatures where the equilibrium yield is relatively low. Figure 6 shows a typical set of results at 1000K and 1 atm, demonstrating that Knudsen regime microporous membranes give smaller yield increases than dense membranes under the same conditions.

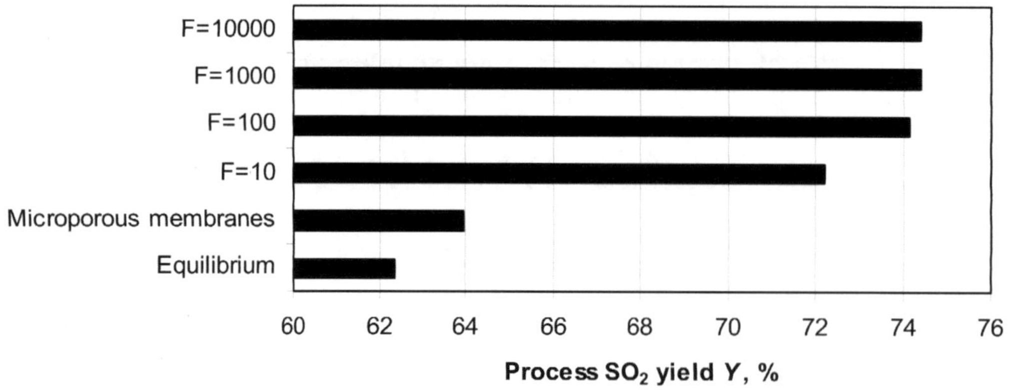

Figure 6: Downstream separation/recycle, 75% membrane O2 permeation, R=80%, T=1000K, feed of 92.7wt% H_2SO_4, P=1 atm

The data in Figure 6 also shows that increasing the selectivity of a dense membrane above 100 has a negligible effect on SO_2 yield. Since dense membranes are inherently gas tight, reduced selectivity might be caused by physical defects in the membrane or poor sealing.

Comparison of membrane separation processes

Comparing Figure 4 and Figure 6, which show simulation data at 1 atm and 1000K for CMR and downstream separation processes respectively, demonstrates that the CMR layout may produce greater SO_2 yield increases than the downstream separation, even for a CMR corresponding to two theoretical equilibrium/separation “stages”. On this basis alone, the CMR seems the better of the two notional layouts. However, fabrication of a CMR may present greater challenges than a downstream separation, in terms of required membrane shape, sealing, etc. Therefore, if a modest increase in yield is acceptable, the simpler construction of a downstream membrane process may be sufficient.

Experimental work: YSZ as a potential membrane material

Experiments were carried out to assess the performance and longevity of yttria-stabilised zirconia (YSZ) when used for oxygen separation in the sulphuric acid decomposition section. YSZ powder (containing 8% yttria) was pressed and sintered to form pellets of approximately 24mm diameter and 2mm thickness. The pellets were then coated with either platinum or gold paste and heated to remove the paste’s organic binder to leave a porous metallic electrode on each face. X-ray diffraction (XRD) and impedance spectroscopy (IS) data were recorded before the pellets were used. Comparing before and after XRD data (collected at ambient conditions) will indicate the presence of any additional phases in the pellet following the SO_2 exposure, whilst the IS data (collected at temperatures in the range 200-500°C) will show any changes in the conductivity of the bulk YSZ itself or of the grain boundaries within the sintered pellet. Any surface layer created on the pellet could also be identified from the IS data.

SO_2 exposure of 8% YSZ/Pt pellet at 900°C

A YSZ-Pt pellet was assembled into a bespoke quartz holder (Figure 7,) with external compression and gold foil gaskets sealing the assembly. The gold foil doubled as the electrical linkage of each pellet electrode to the external electrical circuitry, via the metalwork supporting the pellet holder. The apparatus was assembled inside a Carbolite tubular furnace. As shown in Figure 8, the outlet of the quartzware was connected on the permeate side to a Servomex gas analyser, measuring oxygen concentration, whilst the feed side vented to atmosphere.

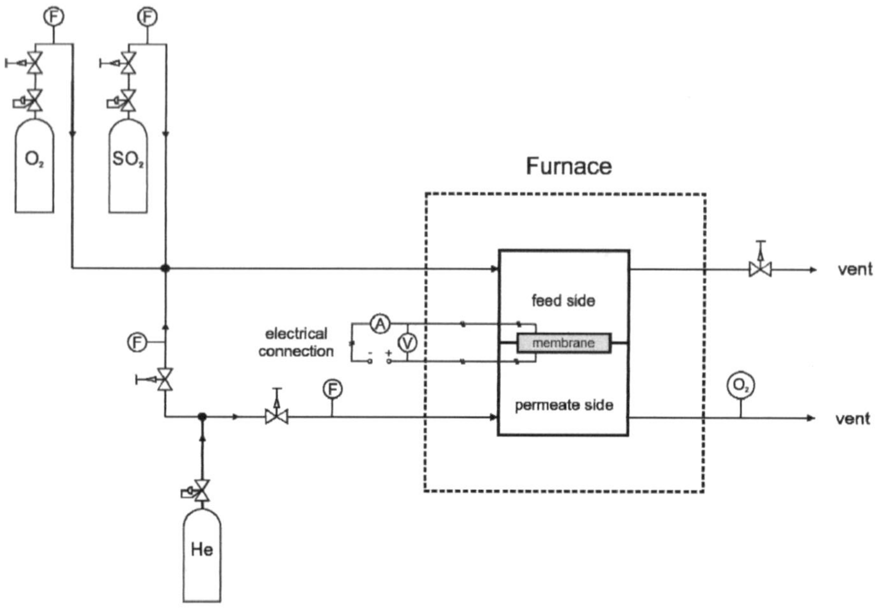

Figure 8: Schematic of experimental apparatus

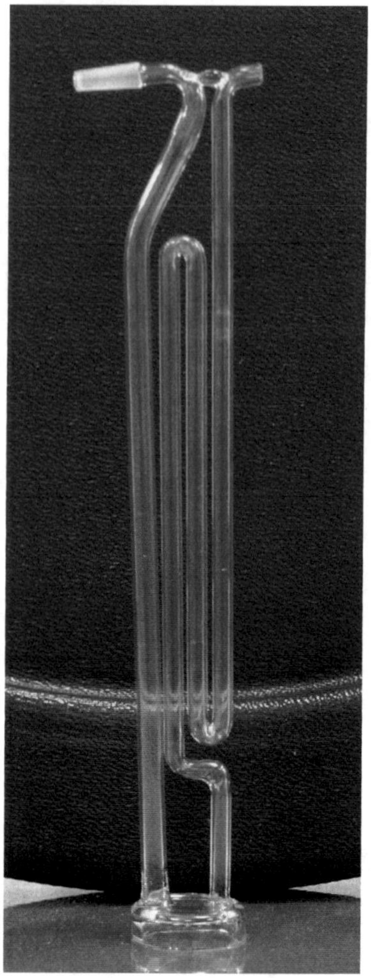

Figure 7: Custom designed quartz membrane holder

The apparatus was heated to 900°C. The feed was 50ml/min of 30% SO_2 and 10% O_2 in He. The permeate side was supplied with 100ml/min of helium. A voltage of 1.00V was applied across the pellet. The current and permeate oxygen concentration were recorded. The feed side of the membrane was exposed to 50ml/min of 30% SO_2 and 10% O_2 in He for 870 minutes over three days. The experiment was ended after 870 minutes as the recorded values had reached a pseudo steady state and it was considered that further exposure time would be of little interest.

At the end of each day's SO_2 exposure period, the permeate oxygen concentration for zero applied voltage was recorded and this value was subtracted from the values recorded that day in order to produce a 'corrected' permeate oxygen concentration representing that oxygen which is being transported through the membrane as a result of the applied voltage and not because of internal electronic

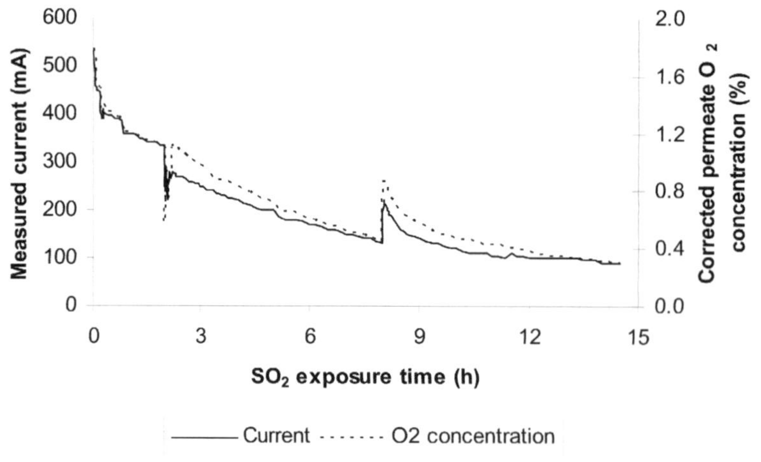

Figure 9: Measured current and corrected measured permeate oxygen concentration for 1.00V applied

transport. The resulting data is shown in Figure 9.

The abrupt changes at 2 and 8 hours signify the start and end points of each day's run. There were some electrical connection problems at the start of the second day, which manifest themselves as the low current values for around two hours. This may also explain the reason why the general trend for the second day is for lower currents compared to days one and three.

Fuming was observed from the feed side outlet during the run. Allowing dampened universal indicator paper to contact the fumes produced a strongly acidic response. Furthermore, a brown liquid collected at the lowest point in the straight quartz cooling tube downstream of the feed side of the membrane. When exposed to air during disassembly, this liquid fumed strongly. These two observations suggest that SO_3 is being produced in the feed side of the experimental apparatus. Given that the reaction rate for the SO_3 decomposition equilibrium reaction is very slow in the absence of catalyst, this indicates that the platinum electrode is catalytically active and is causing a proportion of the supplied SO_2 and O_2 to react and hence move towards the equilibrium position of the SO_3 decomposition step. Further work is in hand to determine the significance of this effect.

SO_2 exposure of 8% YSZ/Pt and 8% YSZ/Au pellets at 800°C

Having investigated the use of an 8% YSZ pellet with Pt electrodes at 900°C, a similar membrane was then tested at 800°C in order that the relative longevity could be measured. Previous equilibrium studies (Figure 10) indicated that zirconia should not suffer sulphation above 600°C. Because of this, a given YSZ membranes exposed to SO_2 at 800°C should not experience a greater percentage decrease in O_2 permeation performance than was recorded at 900°C. An identical YSZ pellet (pressed and sintered at the same time as the YSZ/Pt pellet) but with gold electrodes instead of platinum was tested to compare the O_2 permeation performance and to check if gold electrodes promoted the same catalytic activity seen with platinum electrodes which caused SO_3 to be produced from the SO_2/O_2 feed.

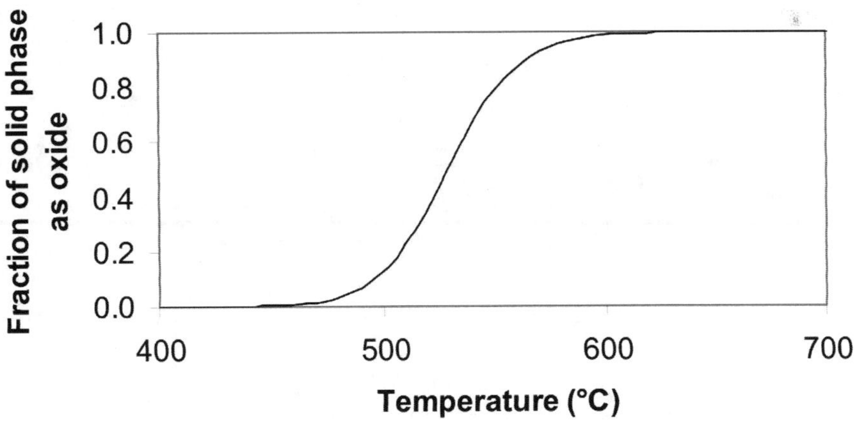

Figure 10: Equilibrium data from *HSC Chemistry* for a solid $ZrO_2/Zr(SO_4)_2$ phase in a gaseous $H_2SO_4/SO_3/SO_2/O_2/H_2O$ phase at 1 bar

The method employed was identical to that used for the 900°C run described previously, except that the ultimate furnace temperature was 800°C not 900°C. Once at 800°C, with 1.00V being applied across the electrode faces, the feed side was supplied with 10% O_2 in

He (total 50ml/min) for one hour. The feed was then changed to 30% SO_2 and 10% O_2 in He (total 50ml/min) for two hours. Between 60 and 90 minutes into this two hour SO_2 exposure period the electrical circuit was opened. This was carried out in order to determine if any increase in the O_2 permeate concentration occurred after a "rest" period of no externally applied voltage. Following the two hour SO_2 exposure period, the feed changed to 10% O_2 in He (total 50ml/min) for one hour. Finally, the feed reverted to 30% SO_2 and 10% O_2 in He (total 50ml/min) for one additional hour, before the experiment was ended.

The measured permeate O_2 concentrations are presented in Figure 11 for the YSZ/Pt membrane and Figure 12 for the YSZ/Au. Additionally, Table 1 shows some of the key comparison data between the platinum and gold electroded membranes.

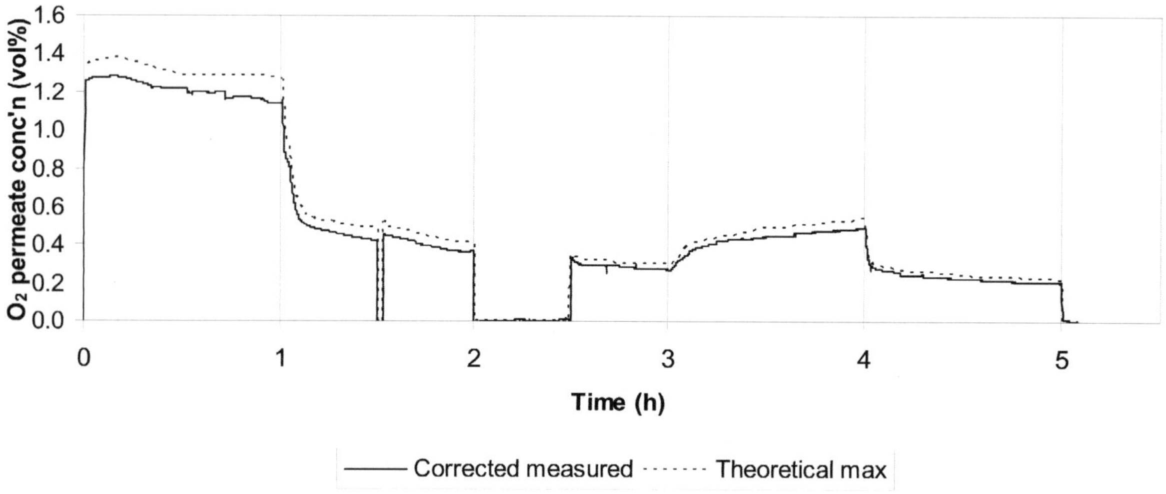

Figure 11: Data for exposure of 8% yttria YSZ pellet with platinum electrodes to SO_2 at 800°C – theoretical O_2 concentration is that calculated from the simultaneous current readings

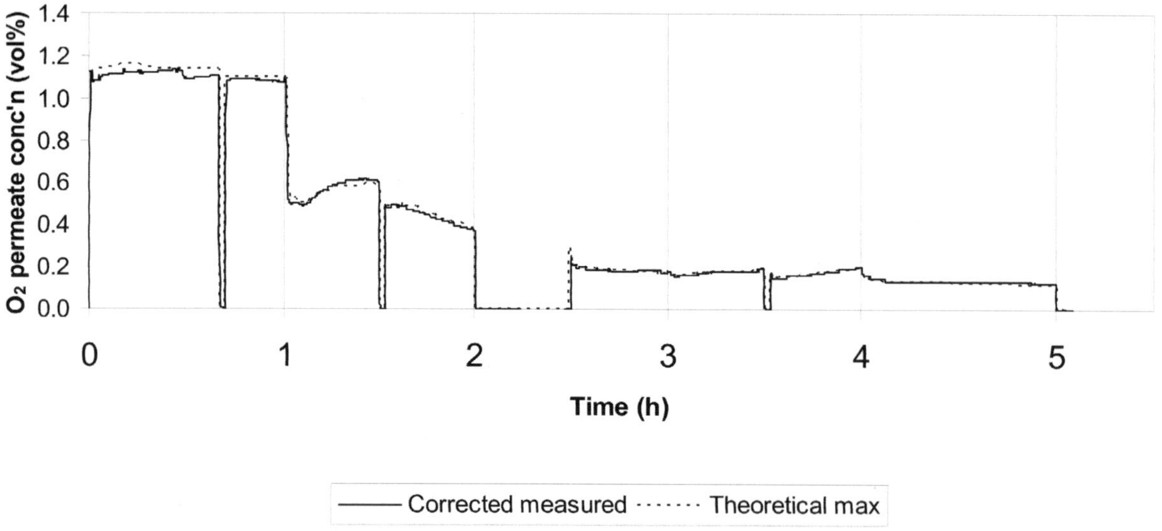

Figure 12: Data for exposure of 8% yttria YSZ pellet with gold electrodes to SO_2 at 800°C – theoretical O_2 concentration is that calculated from the simultaneous current readings

Table 1: Selected data from SO_2 exposure tests of 8% yttria YSZ pellets at 800°C

Permeate O_2 concentration (%) measurement time	Electrode	
	Pt	Au
Initial	1.25	1.13
After 1h O_2-He feed	1.14	1.07
After 10m SO_2-O_2-He feed	0.49	0.54
After 2h SO_2-O_2-He feed	0.26	0.17
After further 1h O_2-He feed	0.49	0.20
After final 1h SO_2-O_2-He feed	0.21	0.12
Ratio after:before further 1h O_2-He	1.88	1.18
Ratio initial:final	5.95	9.42

Table 1 shows the final O_2 concentration to be 1/5.95 of the initial value for the YSZ/Pt pellet at 800°C. This is very similar to the value recorded at 900°C (see Figure 9.) The fact that the fractional reduction in O_2 permeation flux is independent of temperature suggests the absence of a problem with pellet temperature such that the pellet is at a lower temperature than expected and is becoming affected by sulphation (see Figure 10.) Figure 13 shows the combined conductivity of the bulk YSZ in the pellet and the boundaries between the YSZ grains in the sintered mass. There is no significant change in the conductivity following the SO_2 exposure, which indicates that the "bulk" of the pellet is electrically unaffected. This leads to the conclusion that the reduction in permeate O_2 concentration and associated current flow is caused by a surface effect. Other IS and XRD data suggests that the loss of oxygen permeation performance is caused by the creation of a resistive surface layer. Investigations into the nature of this layer are ongoing.

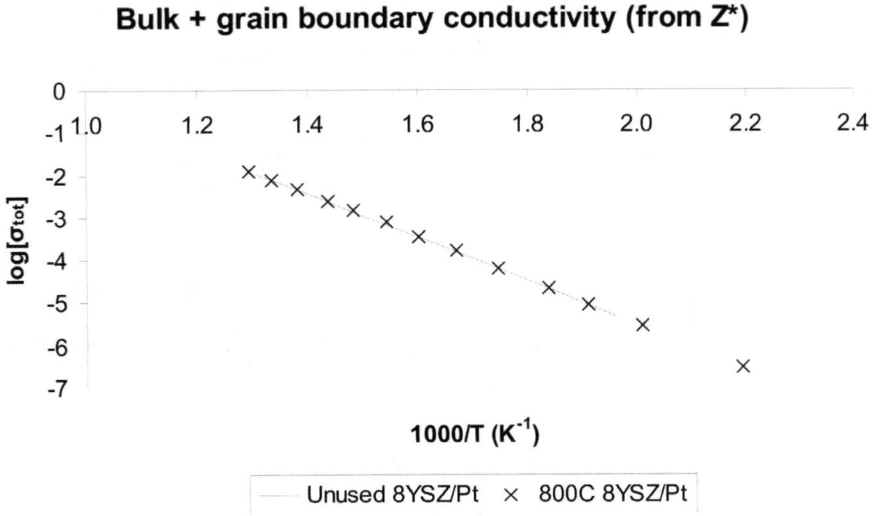

Figure 13: Bulk and grain boundary conductivity against temperature for 8% yttria YSZ/Pt pellet before (line) and after (crosses) SO_2 exposure

A sub-test was performed to establish if the application of the potential across the pellet was responsible for the decreasing permeation performance. Between 2.0 and 2.5 hours into the run, SO_2 and O_2 were supplied to the membrane feed side whilst the electrical circuit was open. As shown in Figure 11 and Figure 12, no oxygen permeated during this period and, following reapplication of the 1.00V potential, no recovery in membrane O_2 permeation performance was observed as a result of the open circuit period for either membrane. That the O_2 concentrations before and after the open circuit period describe a similar path suggests that the degradation in performance is independent of external voltage application or amount of O_2 permeating.

Another sub-test performed was to supply 10% O_2 only (i.e. no SO_2) to the pellets for one hour in order to determine if the pellet would recover some or all of the lost O_2 permeation performance. This was carried out between three and four hours into the run. Within five minutes of reapplying the SO_2-O_2-He feed, the permeate O_2 concentration decreases to that which, extrapolating the pre-recovery values, would be expected at that time. This suggests that replacing the SO_2-O_2-He feed with an O_2-He feed does not extend the membrane life but may merely mask a continuing decrease in performance. The surface effect which is suspected to be causing the performance decrease would appear to be non-reversible through a change in process conditions.

It can be seen from the ratios at the bottom of Table 1 that the YSZ/Au has a greater decrease in O_2 permeation over the course of the run compared to the YSZ/Pt pellet. Additionally, the YSZ/Au only "recovers" by 18% during the one hour O_2-He feed period, compared to the 88% O_2 permeation increase in the YSZ/Pt pellet. This shows that the decrease in O_2 permeation performance is electrode material dependent. The conductivity measurements shown in Figure 14 show that, like the YSZ/Pt pellet, the "bulk" of the pellet has been electrically unaffected by the SO_2 exposure and hence the performance decrease is likely to be related to the surface characteristics.

During both the YSZ/Au and YSZ/Pt pellet investigations at 800°C, fuming indicative of SO_3 production by catalysis at the metallic electrode was observed. The gold electrode therefore also acts catalytically on the SO_2 and O_2 being supplied.

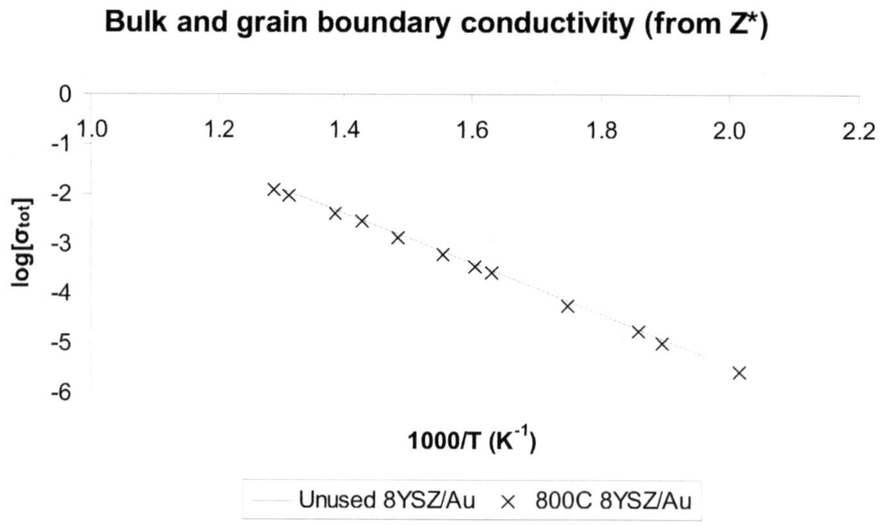

Figure 14: Bulk and grain boundary conductivity against temperature for 8% yttria YSZ/Au pellet before (line) and after (crosses) SO_2 exposure

Conclusions

Process simulations have shown that removing oxygen from the sulphuric acid thermal decomposition process can increase SO_2 yield at a given temperature and pressure. Yttria-stabilised zirconia (YSZ) is a potential dense membrane material for this process. Experiments have shown that whilst the bulk of a YSZ pellet is not affected by SO_2 exposure, a surface effect reduces the O_2 permeation flux at a given applied voltage. Investigations into the nature of this effect are ongoing.

References

[1] Forsberg C, Trowbridge L, Bischoff B, Mansur LK. Sulfur thermochemical processes with inorganic membranes to produce hydrogen. AIChE Spring National Meeting, New Orleans, Louisiana, USA, 2004.

[2] Ozturk IT, Hammache A, Bilgen E. An improved process for H2SO4 decomposition step of the sulfur-iodine cycle. Energy Convers Manage 1995;36(1):11-21.

273e Recent Progress In Catalyst Development for Hydrogen Production In the Sulfur-Iodine Thermochemical Water Splitting Cycle

Daniel M. Ginosar[1], Lucia M. Petkovic[1], Helen H. Farrell[2], and Kyle C. Burch[1]. (1) Chemical Sciences, INL, P.O. Box 1625, Idaho Falls, ID 83415-2208, (2) Material Sciences, INL, P.O. Box 1625, Idaho Falls, ID 83415-2211

Thermochemical cycles produce hydrogen through a series of chemical reactions where the net result is the production of hydrogen and oxygen from water at much lower temperatures than from direct thermal decomposition. All chemicals within the cycle are recycled and the heat to drive the reactions, which tend to be endothermic, must be provided by a primary energy source. When the primary energy driver is nuclear or solar heat, hydrogen can be generated without producing greenhouse gasses, and can provide independence from our dwindling supplies of fossil fuels.

Among the high number of thermochemical water-splitting cycles proposed in the literature, the Sulfur-Iodine (S-I) cycle has generated considerable interest. The S-I cycle consists of three simultaneous reactions; the decomposition of hydroiodic acid (HI) to produce hydrogen and generate iodine for recycle, the decomposition of sulfuric acid to produce oxygen and generate SO2 for recycle, and a main reaction where water and the recycle chemicals react to regenerate HI and sulfuric acid. Both the hydroiodic acid decomposition and the sulfuric acid decomposition reactions require a catalyst to achieve significant rates and product yields at expected operating temperatures.

This paper will focus primarily on catalysts that have been examined for the sulfuric acid decomposition reaction. Activities, stabilities, and material properties changes of several metal oxides, perovskites, and supported platinum group metal catalysts will be presented.

A Simulation Study of Catalytic SO_3 Decomposition in Heat Exchanger Reactor

Shinji Kubo[1], *Nobuyuki Tanaka*[1], *Hirofumi Ohashi*[1], *Akihiro Kanagawa*[2], *Yoshiyuki Imai*[1]
Seiji Kasahara[1] *and Kaoru Onuki*[1]

[1]*Japan Atomic Energy Agency*
Narita-cho 4002, Oarai-machi, Higashi Ibaraki-gun, Ibaraki-ken, Japan

[2]*Mitsubishi Heavy Industries, LTD*
Wadasaki-cho 1-1-1, Hyougo-ku, Kobe-shi, Hyogo-ken, Japan

ABSTRACT

A SO_3 decomposition is common chemical reaction employed by thermochemical water-splitting cycles for hydrogen. This reaction proceeds endothermically in catalytic bed; in heat utilization system of the VHTR, needed heat is provided from high temperature helium gas by heat exchange. To find sizes and shapes in this reactor, a reaction rate of SO_3 decomposition was formulated by a Pt catalyst kinetic measurement experiments; besides, ordinary differential equations for this catalytic bed to produce hydrogen of 1 mol/s, in which the kinetic and heat exchange were counted, were integrated numerically. Characteristics length which represented reactor shapes and required heat transfer area were shown to attain nearly chemical equilibrium. In reference case, it was found that a required heat transfer area was 20 m^2 and very slim flow channels performed the decomposition well to attain near the chemical equilibrium.

INTRODUCTION

A huge demand for hydrogen as an energy carrier is expected for the near future. The thermochemical water-splitting IS processes$^{(1)}$ are due to offer massive hydrogen production methods without carbon dioxide to be supplied with heat from the high-temperature gas-cooled reactors (HTGRs), so that this process attracts broad interests to implement studies. The SO_3 decomposition reactor for oxygen production is particularly important, since large amount of heat is absorbed for the endothermic chemical reaction in a catalytic bed by receiving specific heat of high temperature helium gas. Computing heat exchange type reactor were carry out, in which the kinetics of SO_3 decomposition reaction formulated by a Pt catalyst experiments and heat exchange between the process gas and the helium take place. To investigate achieving nearly chemical equilibrium, required heat transfer area and flow channel shapes of the catalytic bed are discussed.

GOVERNING EQUATIONS

The SO_3 decomposition reaction proceeds endothermically in two stages to produce oxygen. The aqueous phase of sulfuric acid is vaporized, then, decomposed according to following reactions (1, 2):

$$\mathrm{H_2SO_4(g) \longrightarrow H_2O(g) + SO_3(g)} \tag{1}$$

and

$$\mathrm{SO_3(g) \longrightarrow SO_2(g) + \frac{1}{2}O_2(g)}, \tag{2}$$

where "g" means the gaseous phase. Firstly sulfuric acid decomposes spontaneously into sulfur trioxide and gaseous water. Secondly, in higher temperature, the sulfur trioxide decomposes into sulfur dioxide and oxygen, reaction (2), in the presence of solid catalyst.

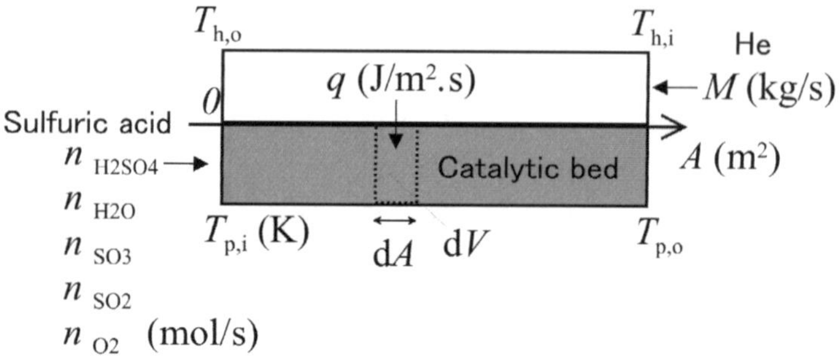

Figure 1: Calculated object of heat exchange type SO_3 decomposition reactor. SO_3 flows in under side channel, helium gas flows in upper side channel. Catalyst completely fills under side channel.

A calculated object of SO_3 decomposition reactor is shown in Figure 1. Heat flux q W/m^2 through the upper side to the under side is given by

$$q = U(T_h - T_p), \tag{3}$$

where U W/m^2·K is overall heat transfer coefficients, T_h K is temperature of the helium gas, T_p is temperature of process gas. With the heat transfer area A m^2, q is expressed in

$$q = M c_{p,h} \frac{dT_h}{dA}, \tag{4}$$

where M kg/s is flow rate of helium gas, $c_{p,h}$ J/kg·K is its specific heat. Here, a characteristic length z m was defined as

$$z = \frac{dV}{dA}, \tag{5}$$

an ordinary differential equation on temperatures:

$$n_{\mathrm{total}} \frac{dH}{dT} \frac{dT_p}{dA} = -\Delta H \cdot z \cdot r + q \tag{6}$$

was obtained, where V is volume of the catalytic bed, n mol/s are constituent flow rates of process gas, n_{total} is total molar flow rate of process gas, H J/mol is molar enthalpy of process gas, ΔH J/mol heat of reaction for reaction (2) and r mol·m^{-3}·s^{-1} is reaction rate of reaction (2). Heat of reaction for reaction (1) was considered by the term of $\mathrm{d}H/\mathrm{d}T$.

PARAMETER SETUP

The parameters in eqs.(3-6) were set up in such a way as to integrate eq. (6).

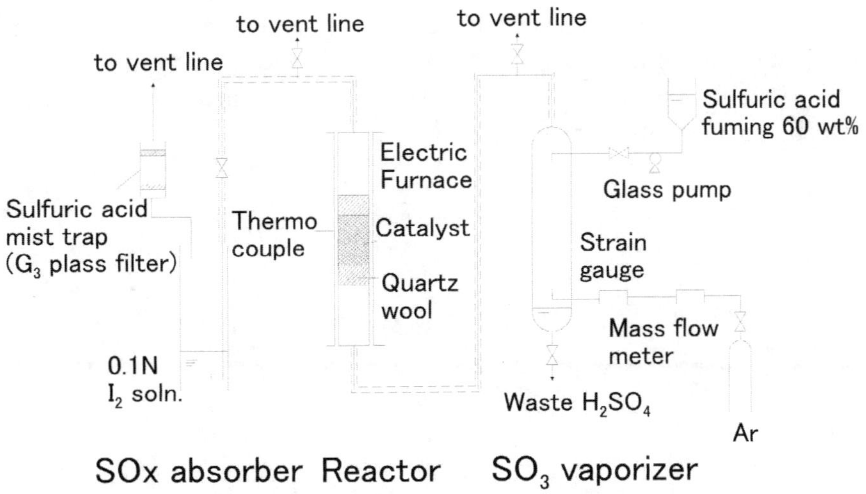

Figure 2: Experimental apparatus(3) for SO_3 decomposition.

Overall heat transfer coefficient U is evaluated from a conceptual design(2) of advanced process heat exchangers. For the SO_3 decomposer, a heat transfer coefficient on helium gas side of 1199 W/m^2·K was specified, another on process gas side of 130 was specified; therefore, 117.3 is employed for U in eq. (3).

A reaction rate equation was formulated from a results of a kinetic measurement experiment(3). Figure 2 shows a schematic of kinetic measurement apparatus. A 1 wt% Pt catalyst on support material of α alumina (Al_2O_3),which bulk density was 1.12 g/cm^3, have been applyed for SO_3 decomposition reaction.

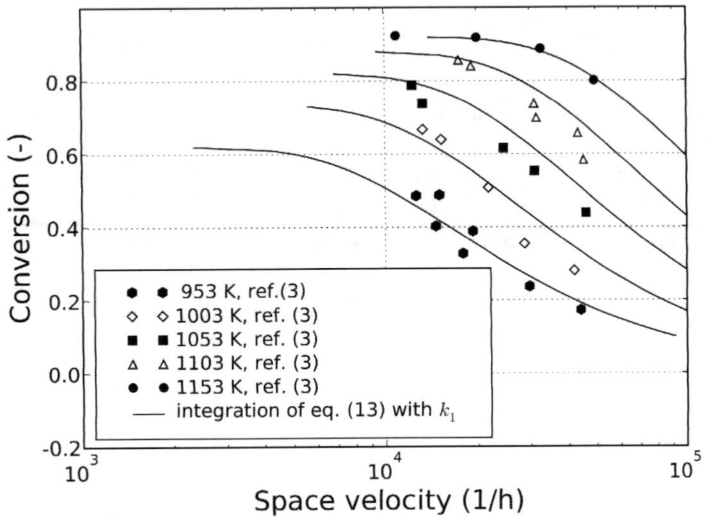

Figure 3: SO_3 decomposition(3) over 1 wt% Pt-αAl_2O_3 catalyst. Initial SO_3 concentration in Ar, 12 mol%.

The experiments have been conducted under atmospheric pressure. SO_3 gas produced from 60 wt% of fuming sulfuric acid was fed to a quartz reaction tube in diameter of 5 mm. The concentration of the feed SO_3 gas was 12 mol%. Space velocity of experimental parameter have been adjusted by flow rate of a carrier gas.

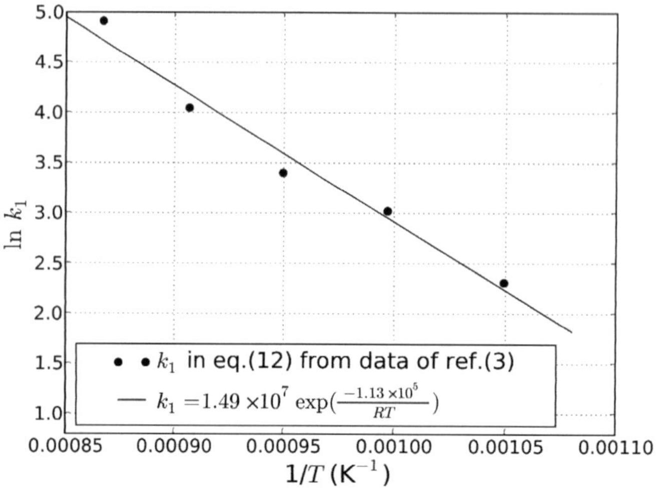

Figure 4: Arrhenius plot of SO_3 decomposition under eq. (14).

Conversion of SO_3 have been measured by chemical titration; the iodine titration was adapted for SO_2 of decomposition product and the neutralization titration was adapted for SO_3 of reactant. Figure 3 illustrates the kinetic of catalytic SO_3 decomposition. As the experimental results shows the Pt catalyst has high catalytic activity, especially in high temperature region of above 1070 K, the decomposition gas attains a chemical equilibrium state on the space velocity of 10000 h^{-1} .

In this study, a reaction rate equation of SO_3 decomposition was formulated from the experimental results of kinetic measurement. Considering that SO_3 decomposition reaction of eq. (2), to begin with, a following reaction rate equation:

$$r = -k_1 C_{SO_3} + k_2 C_{SO_2} C_{O_2}^{0.5} \tag{7}$$

was supposed for primary simple and easy in engineering utilization, where C (mol/m^3) is molar volume concentration, r (mol/m$^3 \cdot$ s) is reaction rate and k is rate constant. Here, supposing a ratio of k_1 to k_2 is constant; with $r = 0$ at equilibrium sate, the ratio was given by

$$\frac{k_2}{k_1} = \frac{C_{SO_3}}{C_{SO_2} C_{O_2}^{0.5}} = (\frac{RT}{P})^{0.5} \frac{X_{SO_3}}{X_{SO_2} X_{O_2}^{0.5}}, \tag{8}$$

where X is molar fractions, P_0 (Pa) is reference pressure, T_0 (K) is reference temperature, P is total pressure, T is reaction temperature and R is the gas constant. The equilibrium constant K is

$$K = (\frac{P}{P_0})^{0.5} \frac{X_{SO_2} {X_{O_2}}^{0.5}}{X_{SO_3}} = \exp(-\frac{\Delta G}{RT}), \tag{9}$$

where ΔG (J/mol) is change of free energy. With K, rewritten eq. (8) is obtained:

$$\frac{k_2}{k_1} = \underbrace{(\frac{RT}{P})^{0.5}(\frac{P}{P_0})^{0.5}}_{B}\frac{1}{K} = B\frac{1}{K}. \tag{10}$$

Therefore, k_2 was given by

$$k_2 = k_1 B\frac{1}{K}. \tag{11}$$

As a consequence, reaction rate r was given by

$$r = k_1(-C_{SO_3} + \frac{B}{K}C_{SO_2}C_{O_2}^{0.5}). \tag{12}$$

Subsequently, the integration method was employed to specify reaction constant k_1. A relationship between space velocity S_v (s^{-1}) and the conversion α is given by

$$\frac{1}{S_v}\frac{PT_0}{TP_0} = \tau = C_0\int_0^{\alpha_x} -\frac{1}{r}d\alpha, \tag{13}$$

where C_0 mol/m^3 is initial molar volume concentration at reaction temperature and pressure and τ (s) is space time. With r, therefore, τ was expressed by

$$\tau = -C_0\int_0^{\alpha_x} r^{-1}d\alpha = -C_0 k_1^{-1}\int_0^{\alpha_x}(-C_{SO_3} + \frac{B}{K}C_{SO_2}C_{O_2}^{0.5})^{-1}d\alpha. \tag{14}$$

Figure 4 shows a Arrhenius plot of the experimental data of kinetic measurement(3) using eq. (14). This figure highlights that eq. (12) can represent acceptably the experimental data. The solid curves in Figure 3 were calculated conversion by using obtained k_1.

CONDITION SETTING

For H_2SO_4 dissociation in eq. (1), chemical equilibrium states were assumed at every location of A.

Minimum temperature difference $\Delta T_{\rm min}$ between process side and helium side was sat at 50 K(4). Production flow rate of SO_2, n_{SO2}, of 1 mol/s was chosen. Temperature at the outlet $T_{\rm p,o}$ was obtained as $T_{\rm h,i}$ - $\Delta T_{\rm min}$. Conversion at outlet was calculated using temperature of $T_{\rm h,i}$ - $\Delta T_{\rm min}$, the conversion was adjusted to value of 0.9 times lower than equilibrium at temperature of $T_{\rm h,i}$ - $\Delta T_{\rm min}$ (near equilibrium state).

Following conditions of inlet sulfuric acid were adopted from flow sheet analyses(5) for H_2SO_4 processing section: Temperature 800 K; Total pressure 1.2 MPa; Concentration of H_2SO_4 92.7 wt%.

Conditions of inlet helium gas were adopted from a conceptual design of HTTR-IS hydrogen production system(6). Temperature at inlet $T_{\rm h,i}$ was adapted at 880+273.15 K. Flow rate of helium gas was calculated by

$$M = n_{\rm so2}\, 285.8 \times 10^3 \frac{1}{\eta}\frac{1}{c_{p,h}(T_{\rm h,i} - T_{\rm h,low})}, \tag{15}$$

where n_{so2} is 1 mol/s, overall thermal efficiency to produce hydrogen is η=0.4 and temperature from total system outlet $T_{\text{h,low}}$ is 526 K.

Data of enthalpies and free energy differences of thermodynamic tables[(7)] were used for the computations.

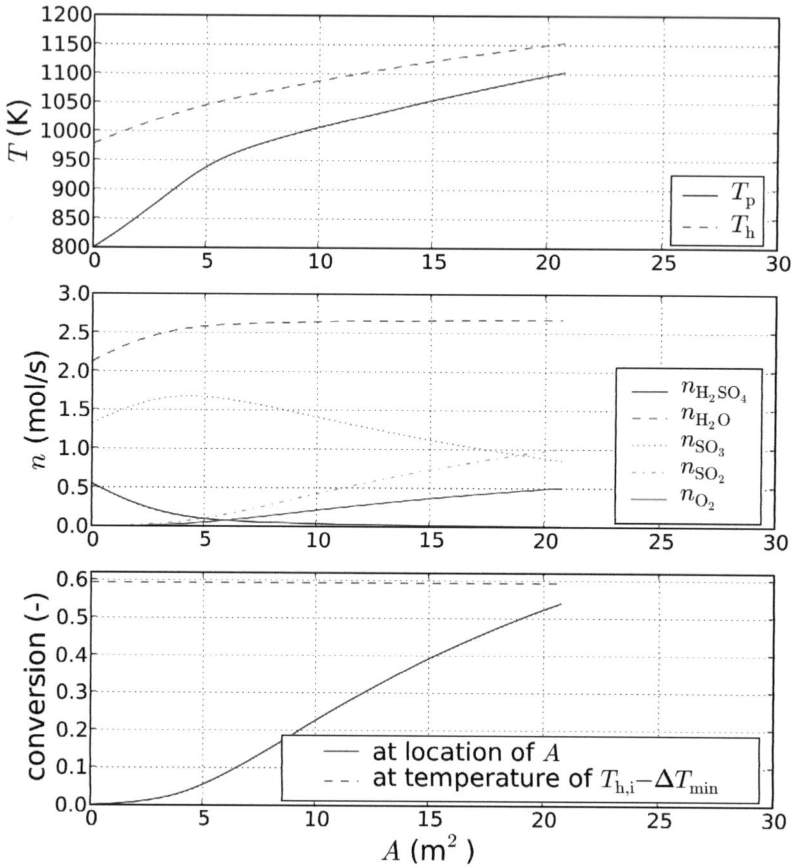

Figure 5: Integration result of eq. (6) under a_r=1.0 and a_U=1.0 on specified conditions (e.g. ΔT_{min}=50 K).

RESULTS AND DISCUSSION

Numerical integrations of eq. (6) were carried out by Euler method. On parametric calculations, following ratios a were employed:

$$r = a_r r_{\text{norm}} \tag{16}$$

and

$$U = a_U U_{\text{norm}}, \tag{17}$$

where r_{norm} is norm value of eq. (12) and U_{norm} is norm value of the 117.3 W/m^2·K. Figure 5 shows distributions of variables in the heat exchange type SO_3 decomposition reactor on a_r=1.0

and a_U=1.0. This figure highlights that the specified conditions (e.g. $\Delta T_{\min}$=50 K) are satisfied in this numerical calculations.

Computations in various a_r and a_U were conducted to find A and z, at which the decomposition gas were near the chemical equilibrium state. Figure 6(a) shows contour map of required area of the reactor. The area are independent of r; in norm case, the area is in 20 - 25 m^2.

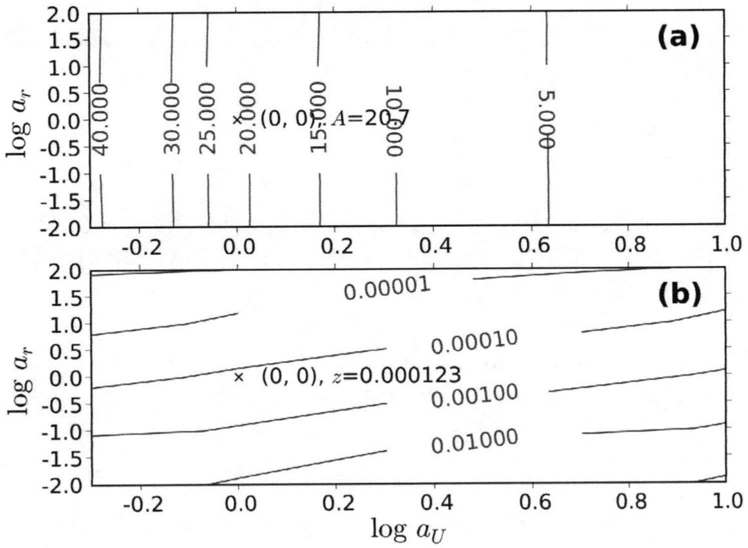

Figure 6: Required heat transfer area A and characteristics length z to attain equilibrium decomposition for heat exchange type SO_3 decomposition reactor. (a) Contour map of required area A. (b) Contour map of characteristics length z in eq. (5). Point of (0, 0) represents norm conditions of a_r=1.0 and a_U=1.0 .

Figure 6(b) shows contour map of the characteristics length z in eq. (5). z represents shape of the catalytic bed, for example, $z = 1$ means diameter of 4 m in case of the circular tube. In case of above z, the decomposition gas could reach near the equilibrium state. In norm case, very slim flow channel even perform the decomposition reaction well to attain near the chemical equilibrium.

SUMMARY

SO_3 decomposition is common chemical reaction employed by thermochemical water-splitting cycles for hydrogen. This reaction proceeds endothermically in catalytic bed; in heat utilization system of the VHTR, needed heat is provided from high temperature helium gas by heat exchange. To specify sizes and shape of this reactor, the kinetic of SO_3 decomposition reaction was formulated by a Pt catalyst experiments for primary simple and easy in engineering utilization; besides, the ordinary differential equation for catalytic bed, in which the kinetic and heat exchange were considered, were computed. The characteristics length z which represented reactor shapes and the required heat transfer area were shown to attain nearly chemical equilibrium.

REFERENCES

1. J.H.Norman, G.E. Besenbruch and D.R. O'Keefe, Thermochemical Water-splitting for Hydrogen Production, Final Report (January 1975-December 1980), GRI-80/0105, March 1981.

2. N. Sakaba, H. Ohashi, H. Sato, T. Hara, R. Kato and K. Kunitomi, Hydrogen Production by High-Temperature Gas-Cooled Reactor; Conceptual Design of Advanced Process Heat Exchangers of the HTTR-IS Hydrogen Production System, Vol. 7, No. 3, P.247 (2008).

3. S. Sato, Y. Ikezoe, T. Suwa, S. Shimizu, H. Nakajima and K. Onuki, Studies on Closed-Cycle Process for Hydrogen Production, V (Progress Report for the F. Y. 1980), JAERI-M 9724, pp. 53-68 (1981).

4. H. Tatsumi, K. Matsuda, Pinch Technology, The Energy Conservation Center, Japan, p. 31 (2002).

5. I. T. Ozturk, A. Hammache and E. Bilgen, An Improved Process for H_2SO_4 Decomposition Step of the Sulfur-Iodine Cycle, Energy Convers. Mgmt Vol. 36, No. 1, p. 15, 1995.

6. Hiroyuki Sato, Shinji Kubo, Nariaki Sakaba, Hirofumi Ohashi, Naoki Sano, Tetsuo Nishihara, Kazuhiko Kunitomi, Conceptual design of the HTTR-IS hydrogen production system -Dynamic simulation code development for the advanced process heat exchangers in the HTTR-IS system, Proceedings of International Conference on Advanced Fuel Cycles and Systems GLOBAL2007, pp.812-819, 9-13 September 2007, Boise, ID, USA (2007).

7. Chase, M. W., 1998. NIST-JANAF Thermochemical Tables, fourth ed. Am. Chem.. Soc. and Am. Inst. of Physics, Washington, D. C., USA.

Influence of Steam Conversion Rate and Cathode Gas Composition on Performance of High-Temperature Steam Electrolysis Cell

Shinichi Makino[1]*, Kazuya Yamada*[1]*, Masatoshi Hodotsuka*[1]*, Kentaro Matsunaga*[1]*,*

Masato Yoshino[1]*, Eiji Hoashi*[1]*, Shigeo Kasai*[2]*, and Seiji Fujiwara*[2]

(1) Power & Industrial Systems Research & Development Center, Toshiba Corporation

(2) Nuclear Energy Systems & Services Div., Toshiba Corporation

ABSTRACT

High-temperature steam electrolysis is one of the most promising methods for producing hydrogen and has the potential do so with high efficiency. Additionally, this hydrogen production process, combined with nuclear energy, emits no carbon dioxide. Toshiba has been developing a high-temperature steam electrolysis system using solid-oxide cells for hydrogen production in a temperature range of 1073 to 1173K. A tubular cell was selected for development from the viewpoint of leak-tightness. The influence of the steam conversion rate and the cathode gas composition on the cell performance was examined. The effect of the steam conversion rate was examined for a cathode gas composition of 50% hydrogen and 50% steam. The cell performance was not affected by a steam conversion rate of up to 63% and gradually decreased as the conversion rate increased. The effect on the cell performance of cathode hydrogen concentrations from 50% to 0% was also examined for a steam conversion rate of 60%. The cell performance was not affected by hydrogen concentrations ranging from 50% to 5% but dropped when the hydrogen concentration was 0%. The cause of this deterioration is considered to be oxidation of nickel at the cathode surface when the nickel contacts steam. As a result, the hydrogen production efficiency is expected to reach a maximum of 53% at a steam conversion rate of 63% and a cathode hydrogen concentration of 5%. The hydrogen production efficiency of a high-temperature electrolysis (HTE) system, connected to a high-temperature gas reactor, should be greater than that of a conventional water electrolysis system. Based on our findings, the HTE system, combined with nuclear power, is one of the most promising methods of producing hydrogen.

INTRODUCTION

A so-called hydrogen society, where primarily hydrogen will be used as an energy source and carbon dioxide will be scarcely discharged to the environment, has been proposed as one of the best measures against global warming. At present, almost all of the available hydrogen is used as raw material in the chemical industry. It is estimated that the future hydrogen demand will be several times larger than at present because of the widespread adoption of fuel cells. The Japanese government plans to reduce CO_2 emission by replacing 5 million gasoline-powered vehicles with fuel-cell-powered vehicles by 2020. As a result, it is estimated that the demand for hydrogen will increase to 40 billion m^3 at 273K, 0.1MPa a year more than the current level. A large amount of hydrogen is commercially produced by steam reforming, using methane, natural gas or LPG, which originate from fossil fuels. This process, however, is not effective against global warming, because it emits large amounts of CO_2.

Hydrogen production based on High Temperature Electrolysis (HTE), using solid oxide fuel cell (SOFC) technology, in combination with nuclear energy emits no carbon dioxide. For the most part, it uses environmentally sound and commonly available materials. [1]

Toshiba has been developing a tubular solid oxide electrolysis cell (SOEC) that uses heat and electricity from a nuclear power plant as an energy source for hydrogen production. Tubular electrolysis cells have a number of advantages over planar-type cells, including sealing performance and thermal stress, and have been developed from both experimental and numerical approaches.[2,3,4]

An assembly unit consisting of fifteen tubular electrolysis cells was developed. The unit was designed to get the hydrogen production rate of 100L/h at 273K, 0.1MPa and 130L/h at 273K, 0.1MPa had been accomplished. [7] In the paper, it was confirmed that the unit, which consisted of cells, seals, electrical isolation, power supplies and so on, worked well. [5,6,7] However, when hydrogen is produced by HTE, it is important to determine how effectively the supplied thermal energy is used. This paper describes the influence of the steam conversion rate and the cathode gas composition on the performance of an HTE cell.

PRINCIPLE OF HIGH TEMPERATURE ELECTROLYSIS

Figure 1 shows the principle of high-temperature electrolysis (HTE). The electrolysis cell consists of an electrolyte layer between a cathode (hydrogen electrode) and an anode (oxygen electrode) on either side. The electrolyte is made of solid oxide with high oxygen ion conductivity. Supplied steam is decomposed into hydrogen and oxygen ions at the cathode. The oxygen ions then pass through the electrolyte to the anode, where they release electrons and become oxygen. Hydrogen is produced from the supplied steam at the cathode as described in Equation (1) below, and oxygen is released at the anode as described in Equation (2). As a whole, water, in the form of steam here, is decomposed into hydrogen and oxygen as described in Equation (3).

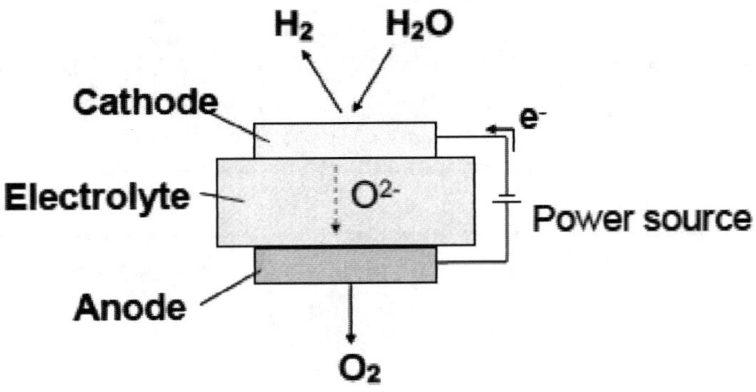

Figure 1 Principle of high-temperature electrolysis (HTE)

$$H_2O + 2e^- \rightarrow H_2 + O^{2-} \qquad (1)$$

$$O^{2-} \rightarrow 1/2\ O_2 + 2e^- \qquad (2)$$

$$H_2O \rightarrow H_2 + 1/2\ O_2 \qquad (3)$$

The required energy for water decomposition is given by:

$$\Delta H = \Delta G + T\Delta S \qquad (4)$$

where ΔH is the enthalpy change, which is equal in magnitude and opposite in sign to the hydrogen heat of combustion, ΔG is the Gibbs free energy change, ΔS is the entropy change in the reaction described in Equation (3), and T is the reaction temperature. In electrolysis, ΔG is given by electrical energy and $T\Delta S$ is given by thermal energy. Figure 2 shows the required energy for electrolysis as a function of temperature. As shown, ΔG decreases with temperature, whereas ΔH hardly changes with temperature. HTE operates effectively at 1073 to 1173 K. HTE basically requires less electrical energy compared with other electrolysis methods carried out at lower temperature.

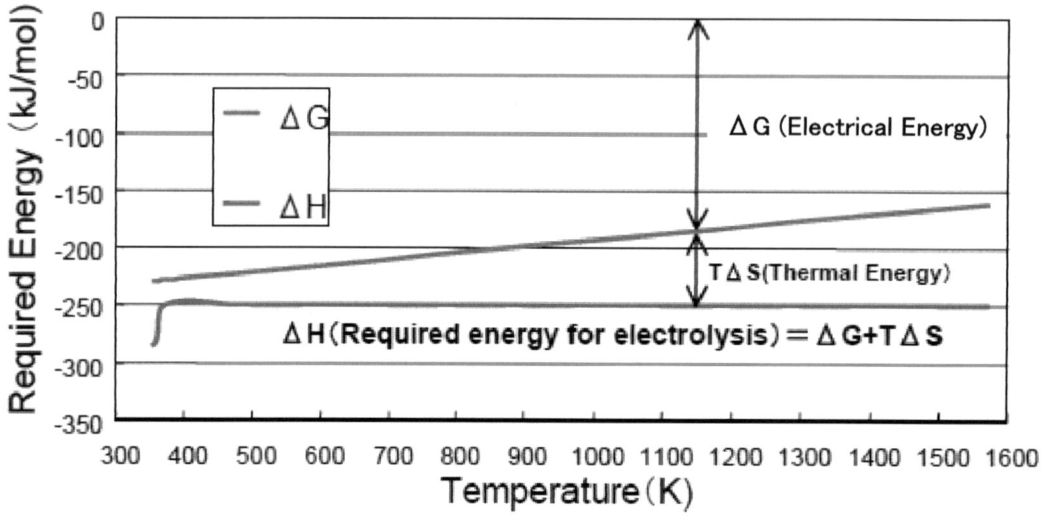

Figure 2 Required Energy for Electrolysis

TUBULAR ELECTROLYSIS CELL

A high-temperature steam electrolysis cell also needs to have thermal, chemical, and mechanical stability to endure long-term operation at high temperature in a red/ox atmosphere. Tubular cells were used because of their superior leak-tightness.

The structure of the tubular electrolysis cell is shown in Figure 3. The cell consists of three layers. The hydrogen electrode was made of nickel-yttrium-stabilized zirconia cermet (Ni-YSZ). The electrolytes was made of yttrium-stabilized zirconia (YSZ). The oxygen electrode was made of mixed oxide of lanthanum, strontium and cobalt. The electrolysis cell was supported by the hydrogen electrode. Steam mixed with hydrogen is supplied to the hydrogen electrodes through a steam introduction pipe provided inside the electrolysis cell, and oxygen mixed with nitrogen is supplied to the oxygen electrodes.

Figure 4 shows a prototype tubular single cell fabricated for lab-scale tests. The outer diameter of the cell is 12 mm. In this cell, the electrolyte is 13 μ m thick, the oxygen electrode is 25 μ m thick, and the hydrogen electrode is 7 μ m thick.

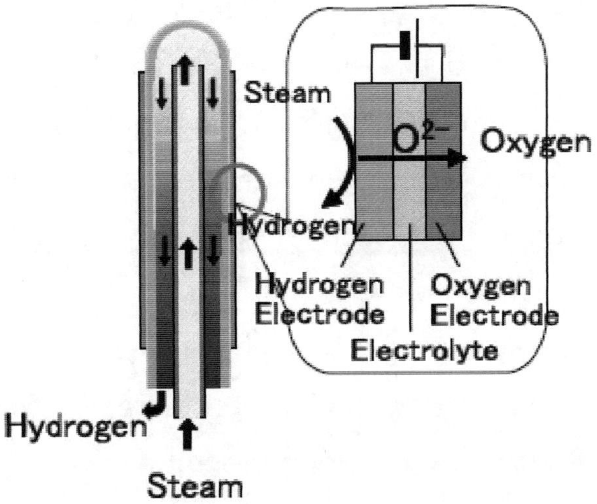

Figure 3 Structure of Tubular Electrolysis Cell

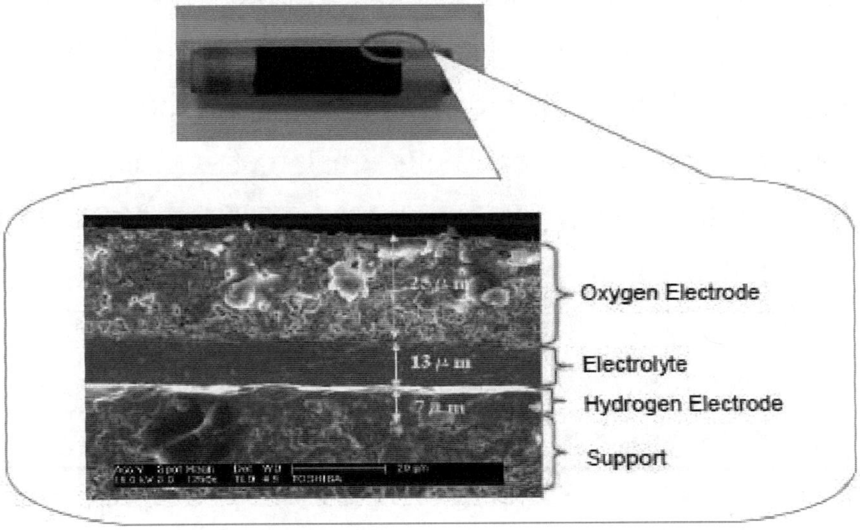

Figure 4 Prototype Tubular Single Cell

HYDROGEN PRODUCTION EFFICIENCY

An important parameter in the design of a hydrogen production system is the hydrogen production efficiency. The power generation efficiency of a High Temperature Gas Reactor (HTGR) is about 50%.[8] When water is electrolyzed with electricity generated by the HTGR, the hydrogen production efficiency is calculated by multiplying the power generation efficiency of the HTGR by the efficiency of the electrolysis. Thus, when the electrolysis efficiency of the water is assumed to be 80%[9], the hydrogen production efficiency is 40%. The hydrogen production efficiency using an HTE must be more than 40%.

An HTE-based hydrogen production system is shown in Figure 5. The inlet of this system is liquid water. The water is evaporated by and its temperature controlled by a preheater. The steam, that is, evaporated water, is fed along the surface of the cathode and decomposes to hydrogen and oxygen. In order to maintain high catalytic activity, the nickel on the surface of the cathode must be in a reduced state. A part of the produced hydrogen is mixed with the supplied steam and is consumed to keep the nickel reduced. The hydrogen production efficiency (ψ) of the HTE system is defined as shown in Equation (5). It is derived by dividing the higher heating value (HHV) of the produced hydrogen by the consumed energy, which is the total of thermal and electrical energy. The thermal energy consists of E_1 and E_2, where E_1 is the thermal energy supplied to the evaporator and E_2 is the thermal energy supplied to the preheater. E_1 is a function of the flow rate of supplied steam, Q_1, as defined in Equation (7). E_2 is a function of Q_1 and the flow rate of circulated hydrogen, Q_2, as defined in Equation (8). The steam conversion rate, α, is defined as the amount of steam converted into hydrogen divided by the flow rate of supplied steam, as given by Equation (6). Therefore, the steam conversion rate is a function of Q_1. The hydrogen concentration supplied to the cathode is a function of Q_2, as defined in Equation (9). Therefore, E_{th} is a function of the steam conversion rate and the hydrogen concentration supplied to the cathode. The hydrogen production efficiency (ψ) of the HTE system is thus a function of the steam conversion rate and the hydrogen concentration. The factors which have an influence on the hydrogen production efficiency are the steam conversion rate and the hydrogen concentration. When there is little supplied steam relative to the flow rate of hydrogen production, the steam conversion rate is high. The thermal energy (E_1) supplied to the evaporator and the thermal energy (E_2) supplied to the preheater decrease, and the total thermal energy (E_{th}) thus decreases. Therefore, the hydrogen production efficiency (ψ) becomes higher. Improving the steam conversion rate causes a decrease in the flow rate of waste of the steam. In addition, the thermal energy (E_{th}) decreases, and the hydrogen production efficiency improves. Once the flow rate of circulated hydrogen (Q_2) is decreased, the hydrogen concentration decreases. And then E_2 decreases, and finally E_{th} decreases. Therefore, the hydrogen production efficiency increases. The hydrogen concentration is defined as the flow rate of circulated hydrogen in the supplied steam.

$$\psi = \frac{HHV}{(E_{el}/\eta)+E_{th}} \quad (5)$$

$$\alpha=Q_4/Q_1 \quad (6)$$

$$E_1=K_1Q_1 \quad (7)$$

$$E_2=K_2Q_1+K_3Q_2 \quad (8)$$

$$E_{th}=E_1+E_2$$

$$C=f(Q_2) \quad (9)$$

$$\phi=F(\alpha,C)$$

ϕ (-): Hydrogen production efficiency

HHV (kJ): Higher heating value

E_{el} (kW/h): Supplied electrical energy

η (-): Generation efficiency

E_{th} (kJ): Supplied thermal energy

E_1: Supplied thermal energy to evaporator

E_2: Supplied thermal energy to preheater

α: Steam conversion rate

K_1, K_2, K_3: Constants

Q_1: Flow rate of supplied steam

Q_2: Flow rate of circulated hydrogen

Q_3: Flow rate of hydrogen produced

Q_4: Flow rate of the steam converted into hydrogen

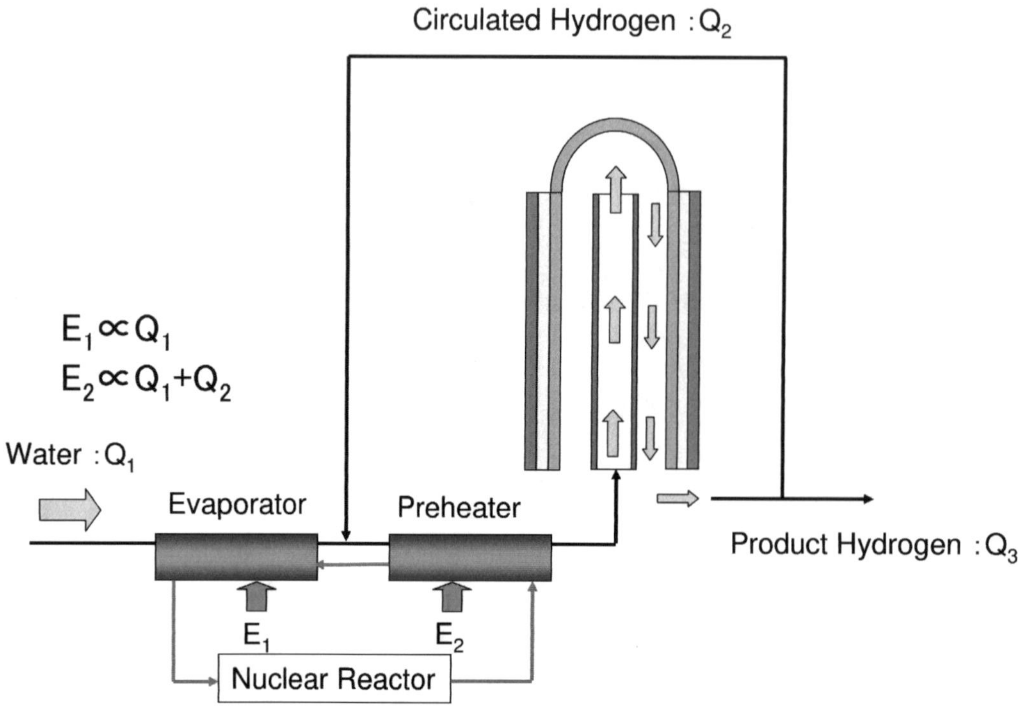

Figure 5 HTE System

APPARATUS

A schematic drawing of the apparatus is shown in Figure 6. The cell is installed inside an electric furnace to keep the cell at constant temperature. The outside of the cell is the anode, and the inside is the cathode. The flow rate of water is controlled with a pump. Water is supplied to the evaporator where it changes into steam, which is then supplied to the cathode. Hydrogen for reduction is supplied together with the steam so that the cathode is not oxidized. The flow rate of hydrogen is measured with a mass-flow controller (MFC). Oxygen and nitrogen are supplied to the anode. The flow rates of the oxygen and nitrogen are measured with MFCs. The flow rate of the exit gas of the cathode is measured with a soap film-type flowmeter, and the exit gas passes through a cooler to remove surplus water. The exit gas of the anode is exhausted after cooling in a cooler.

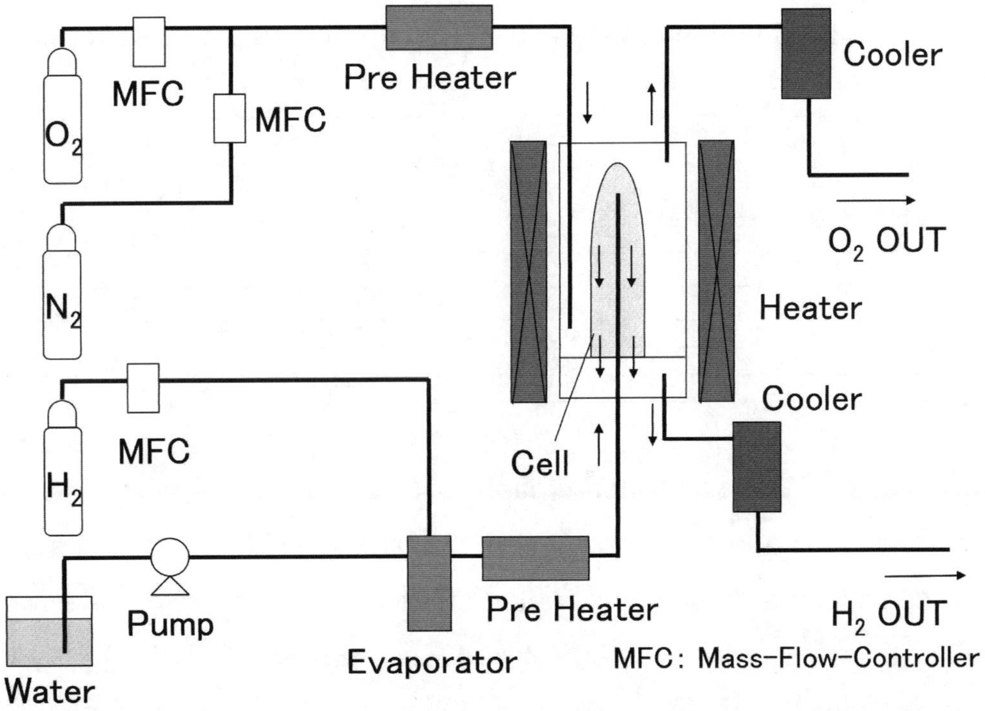

Figure 6 Test Apparatus

TEST CONDITIONS

The evaluation metric for the cell performance was the current density at a cell voltage of 1.3 V at 1073K. The current density was calculated by dividing the measured electric current by the surface area of the electrode. When the electric current supplied to the cell is constant, the hydrogen production rate must be constant. When the hydrogen production rate is constant, the steam consumption is constant. Therefore, the steam conversion rate changes according to the amount of supplied steam. The flow rate of the steam was set as a test parameter for the steam conversion rate. Table 1 shows the test conditions. The gas composition supplied to the cathode was 50% steam and 50% hydrogen. The test parameter was the flow rate of the total amount of gas supplied to the cathode, and the ratio of steam and hydrogen was constant The composition of the gas supplied to the anode was 20% oxygen and 80% nitrogen, and the flow rate of the total amount of gas supplied to the anode was the same as that for the cathode.

Table 1 Test conditions for Influence of steam conversion rate on cell performance

Cell Temp (K)	Cathode		Anode		Cell Voltage (V)
	Steam (L/min) (at 273K, 0.1MPa)	Hydrogen (L/min) (at 273K, 0.1MPa)	Nitrogen (L/min) (at 273K, 0.1MPa)	Oxygen (L/min) (at 273K, 0.1MPa)	
1073	0.09	0.09	0.14	0.03	1.3
1073	0.07	0.07	0.12	0.03	1.3
1073	0.06	0.06	0.10	0.02	1.3
1073	0.05	0.05	0.08	0.02	1.3
1073	0.04	0.04	0.06	0.01	1.3
1073	0.02	0.02	0.04	0.01	1.3

The influence of the hydrogen concentration for reduction on the performance of the cell was examined as follows. The test parameter was the hydrogen concentration supplied to the cathode. The test conditions are shown in Table 2. The steam conversion rate was 20% and 60%. The cathode gas composition was 50% to 100% steam and 50% to 0% hydrogen. The composition of the gas supplied to the anode was 20% oxygen and 80% nitrogen, and the flow rate was the same as that of the gas supplied to the cathode.

Table 2 Test conditions for hydrogen concentration supplied to the cathode

Cell Temp.	1073K
Cell Voltage	1.3 V
Steam conversion rate	20 %, 60 %
Cathode H_2/H_2O	0 %/100 % ~ 50 %/50 %
Anode O_2/N_2	20 % /80 %

RESULTS AND DISCUSSION

Figure 7 shows the test results for the influence of the steam conversion rate on cell performance. The vertical axis shows the normalized cell performance, and the horizontal axis shows steam conversion rate. The cell performance was normalized by the current density of the cell at a cell temperature of 1073K, a cell voltage of 1.3V, a steam conversion rate of 20%, and cathode gas composition of 50% steam and 50% hydrogen. The steam conversion rate was changed by varying the flow rate of the steam. The larger the flow rate is, the smaller the steam conversion rate is. With one cell, several cell performance values can be obtained by varying the steam conversion rate. From the results obtained from four cells, the measurement method is considered to be reproducible. According to the test results, the cell performance was not influenced by the steam conversion rate in the range of 20% to 65%. When the steam conversion rate was more than 65%, the cell performance was considered to deteriorate because of insufficient decomposed steam at the cathode surface.

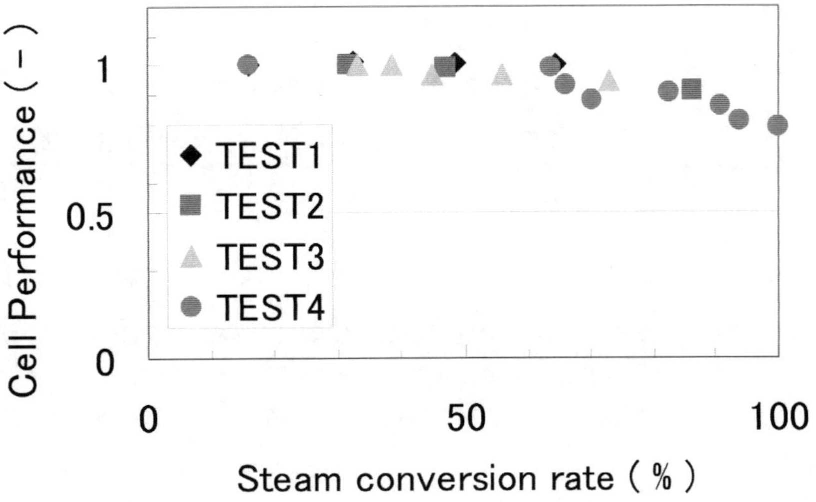

Figure 7 Influence of steam conversion rate on cell performance

Figure 8 shows the influence on cell performance of the concentration of hydrogen supplied to the cathode for keeping it reduced. The vertical axis shows the cell performance, and the horizontal axis shows the hydrogen concentration. The cell performance was not influenced by the hydrogen concentration in the range from 50% to 5%, but the cell performance dropped when the hydrogen concentration was 0%. The cause of this deterioration is considered to be the oxidation of nickel on the cathode surface when the nickel contacts the hydrogenless steam, which is a strong reductant.

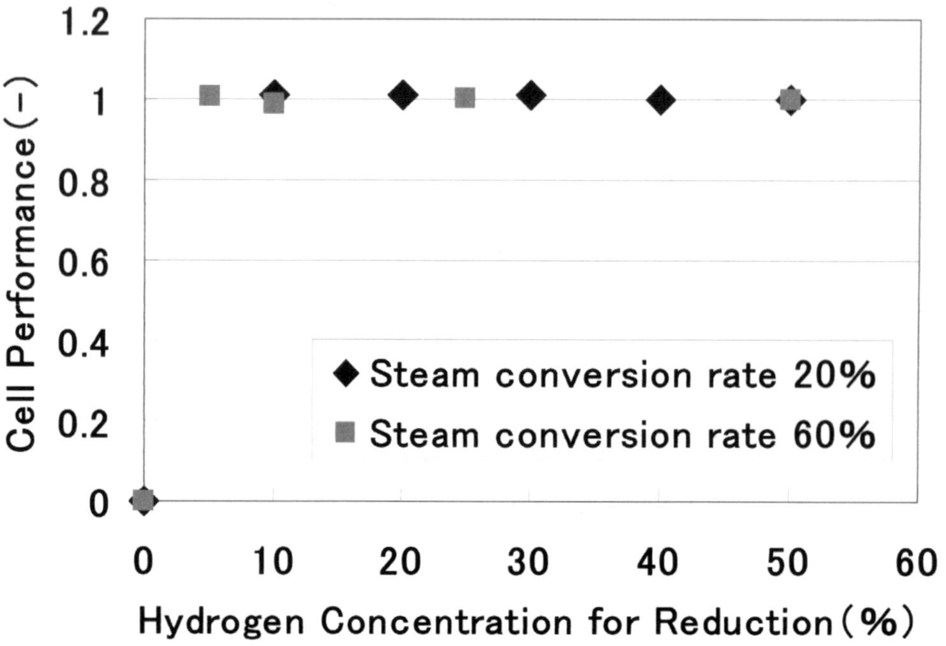

Figure 8 Influence of hydrogen concentration for reduction

The hydrogen concentration for reduction did not affect the cell performance when the hydrogen concentration was more than 5%. The hydrogen production efficiency, given by Equation (5), is the ratio of the higher heating value (HHV) of the product hydrogen to the supplied energy, which is the total of the supplied electrical and thermal energies. The electrical energy is the electrical energy supplied to the cell at a cell voltage of 1.3 V. The thermal energy is the total of the energy supplied to the evaporator and the preheater. The heat loss in the heat exchange was considered to be 0. Figure 9 shows the influence of the hydrogen concentration supplied to the cathode versus the hydrogen production efficiency. The solid lines are the theoretical hydrogen production efficiency given by Equation (5) at a current density of 0.3 A/cm^2. The points are the measured data.

Hydrogen production efficiency reached a maximum of 53% when the steam conversion rate was 60% and the hydrogen concentration was 5%. Steam containing more than 5% hydrogen must be supplied to prevent the cell from oxidizing.

The influence of the hydrogen concentration at the steam conversion rate of 60% was the same as that at the steam conversion rate of 20%. As a result, the hydrogen production efficiency is expected to be maximum at a steam conversion rate of 63% and a cathode hydrogen concentration of 5%.

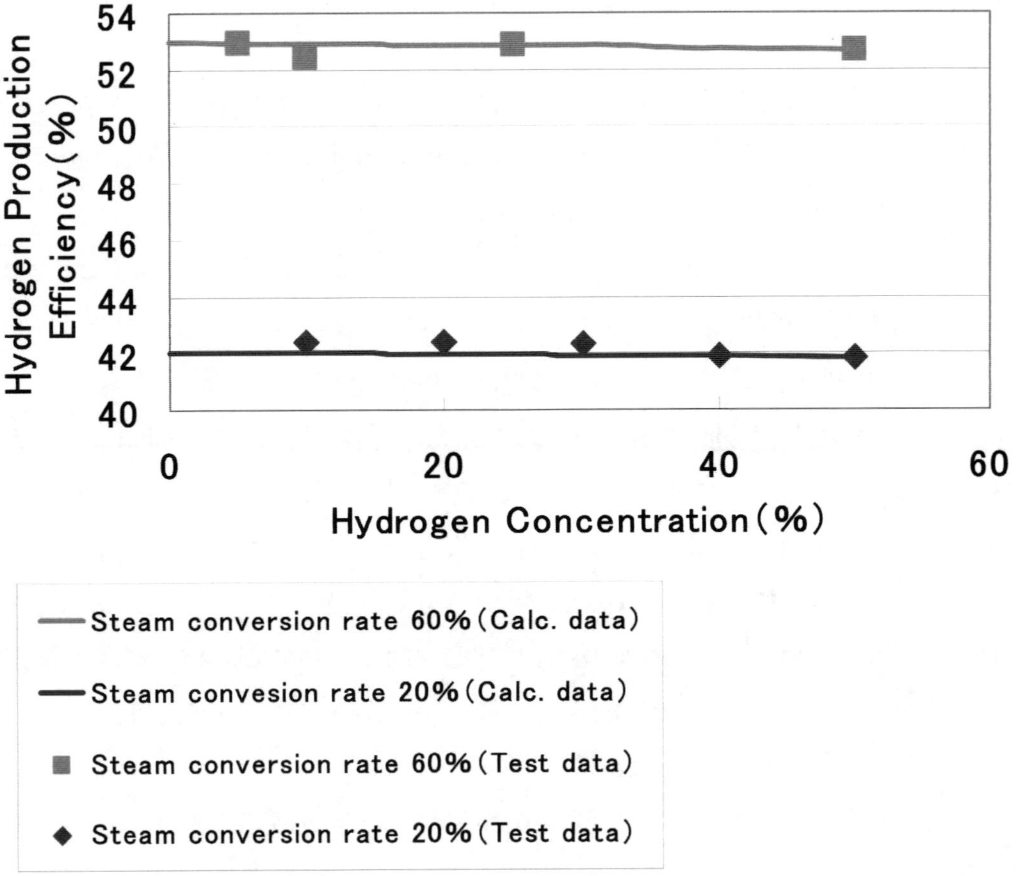

Figure 9 Hydrogen production efficiency

SUMMARY AND CONCLUSIONS

The influence of the steam conversion rate was tested, with the flow rate of the cathode gas as a test parameter. The cell performance was not influenced by the steam conversion rate in the range of 20% to 65%. The influence of the hydrogen concentration for reduction on cell performance was also tested, with the hydrogen concentration as a test parameter. The cell performance was not influenced by the hydrogen concentration supplied to the cathode in the range of 50% to 5%.

As a result, it is expected that the hydrogen production efficiency is maximized at a steam conversion rate of 63% and a cathode hydrogen concentration of 5%. The hydrogen production efficiency of an HTE system, connected to an HTGR, should be greater than that of a conventional water electrolysis system. Based on our findings, an HTE system, combined with nuclear power, appears to be one of the most promising methods of producing hydrogen.

REFERENCES

1. J. S. Herring, J. E. O' Brien, C. M. Stoots, and G. L. Hawkes, " Progress in High-Temperature Electrolysis for Hydrogen Production using Planar SOFC Technology," 2005 AIChE Spring Annual Meeting, April 10 – 14, 2005, Atlanta, GA.
2. N. Osada, H. Uchida, M. Watanabe, N. Suzuki, K. Matsunaga, and M. Yoshino, " High Temperature Steam Electrolysis Cells Using LaCoO3-Based Anode and Ceria-Based Cathode" , Proc. of 206th Meeting of The Electrochemical Society (ECS).
3. K. Matsunaga, E. Hoashi, S. Fujiwara, M. Yoshino, T. Ogawa, and S. Kasai, " Hydrogen Production System with High Temperature Electrolysis for Nuclear Power Plant" , Proceedings of ICAPP ' 06, Paper 6282, Reno, NV USA, June 4-8, 2006.
4. T. Ogawa, K. Matsunaga, M. Yoshino, S. Makino, K. Ono, K. Yamada, S. Kasai, S. Fujiwara, H. Yamauchi, and W. Shinohara, " Steam Electrolysis with Solid Oxide Electrolyte" , Proc. of the Power and Energy Conference of the Institute of Electrical Engineers of Japan, 203 (2006) (in Japanese).
5. K. Yamada, S. Makino, K. Ono, K. Matsunaga, M. Yoshino, T. Ogawa, S. Kasai, S. Fujiwara, and H. Yamauchi, " Development of High Temperature Electrolysis (5), Hydrogen Production using Tubular Electrolysis Cells Integrated Unit" , Proc. of the Fall Meeting of the Atomic Energy Society of Japan, L66 (2006) (in Japanese).
6. A. Ozaki, K. Kubota, and K. Yamada, " Nuclear Hydrogen Production Systems" , Toshiba Review, 60, 2, 27-30 (2005) (in Japanese).
7. K. Yamada, S. Makino, K. Ono, K. Matsunaga, M. Yoshino, T. Ogawa, S. Kasai, S. Fujiwara, and Hiroyuki Yamauchi, " High Temperature Electrolysis for Hydrogen Production using Solid Oxide Electrolyte Tubular Cells Assembly Unit" 2006 AIChE Annual Meeting, Nov.12 – 17, 2006.
8. S. Takada et al. " Design Study of Power Conversion System for the Gas Turbine High Temperature Reactor (GTHTR300)" Proc. of the Atomic Energy Society of Japan Vol. 1, No. 4 (2002) (in Japanese).
9. Agency of Natural Resources and Energy " 1999/2000 Resources energy yearbook" , Investigation Committee for International Trade and Industry Document (1999) p.729.

328b Low Temperature Production of SO2 from H2SO4 In the Si Thermochemical Cycle by Using Iron(III) Sulfate Intermediate

Salvatore Sau, Pietro Tarquini, Pier Paolo Prosini, Alberto Giaconia, Giampaolo Caputo, Raffaele Liberatore, and Michela Lanchi. ENEA, Rome, 00060, Italy

Decomposition of sulphuric acid is the highest temperature step in the sulphur iodine water splitting thermochemical cycle. We propose a variation based on the following reactions:

3 H2SO4(acq,conc) + Fe2O3 -> Fe2(SO4)3 (300 °C)

Fe2(SO4)3 -> Fe2O3 + x SO2 + ySO3 + y/2 O2 (650 °C, where x+y=3)

In comparison with direct sulphuric acid decomposition the maximum temperature is lowered, there is no more need for catalysts and the material costs for the reactor is drastically reduced. One disadvantage is the increase of energy demand.

Experimental and theoretical modeling results are presented and discussed.

A Systematic Methodology for the Evaluation of Alternative Thermochemical Cycles for Hydrogen Production

Ryan J. Andress, Lealon L. Martin, and B. Wayne Bequette
Rensselaer Polytechnic Institute, Troy, NY

1. Introduction

Hydrogen is considered to be an environmentally attractive potential alternative to fossil based fuels. It is an excellent energy carrier, particularly as a carrier of remotely generated electrical energy or as a carrier of residual waste thermal energy. Hydrogen based combustion processes are more efficient than for fossil fuels, releasing 117.6 MJ/kg, compared to only 46.36 MJ/kg for the combustion of n-octane [1]. Hydrogen is also a clean energy source. Its combustion produces no carbon or sulfur products, with water as the primary byproduct. These features characterize an ideal sustainable energy source that can be used to meet increasing residential, commercial, and industrial demands, however, there are several challenges and limitations associated with the production of hydrogen.

Electrolysis of water, a widely used method for hydrogen production, is very inefficient and involves the conversion of primary energy to electricity and electricity to hydrogen [2]. Electrolyzers have been around for centuries [2] with overall process efficiencies between 27% and 32% [3]. Electrolysis requires 140 MJ/kg hydrogen, followed by an additional 15 to 40MJ/kg to compress or liquefy the hydrogen for proper storage and transport [3]. Other methods include fossil fuel reforming, high temperature cracking paired with a high temperature nuclear energy source, and the gasification of biomass and waste [3]. Each of these methods posses distinct challenges and limitations.

Reforming fossil fuels produces low hydrogen concentrations of 70-80%, when catalytic steam reforming is used, and 40-50% for the partial oxidation and autothermal reforming on a dry basis [4]. Although catalytic steam reforming can yield 80% hydrogen it is an energy intensive process and often produces carbon monoxide [4]. The gasification of biomass, such as used edible oils [5], is another option for hydrogen production. Edible oil wastes (EOW) can be blended with high-rank coal (less than 4% moisture when air dried) and gasified to produce fuel. The gasification produces low NO_x and SO_x emissions, however the EOW is 78% carbon and so hydrogen production is limited [5].

The above obtuse production methods arise from the primary challenge with hydrogen production from water, which is the following reaction:

$$2\mathrm{H}_2O_{(l)} \rightarrow O_{2(g)} + 2H_{2(g)} \qquad \Delta G^{\circ}_{298K} = 237 \text{ kJ/mol } \mathrm{O}_{2(g)} \qquad (1)$$

For a reaction to occur the change in Gibbs free energy (ΔG^o) must be negative, therefore reaction (1) does not occur at normal temperatures. Reaction (1) will only occur at temperatures greatly exceeding 2000 K [6], which results in very energy intensive processes to generate hydrogen from water. To overcome this energy barrier a sequence of thermodynamically feasible reactions, whose sum is the overall reaction of splitting water (1), is developed based on non-toxic, abundant compounds. This sequence is referred to as an thermochemical cycle. A representation of a thermochemical cycle is given by the figure below.

Figure 1. Representation of a thermochemical cycle for the production of hydrogen

$$A + B \rightarrow C + D \qquad \Delta G < 0 \quad \forall T \le T_1 \qquad (2)$$

$$D + E \rightarrow A + F \qquad \Delta G < 0 \quad \forall T \ge T_2 \qquad (3)$$

$$B + E \rightarrow C + F \qquad \Delta G > 0 \quad \forall T \qquad (4)$$

The above thermochemical cycle, equations (2) – (4), accomplishes the thermodynamically infeasible ($\Delta G > 0$) reaction (3), with a sequence of alternative reactions, (1) and (2), which are feasible at certain temperatures as shown by Figure 2. In chemical process sequencing, species B and E are consumed producing C and F, while species A and D are recycled. Energy supplied to a thermochemical cycle is largely consumed by the regeneration of the reactants, rather than the formation of the product [1].

The above scheme can be used as a strategy for the splitting of water into hydrogen and oxygen through multiple, feasible, reaction steps at considerably lower temperatures. In such a proposed cycle water is consumed along side other reagents, producing hydrogen and oxygen, while the remaining products are recycled back as reactants. The net reaction for the cycle is (1). The appeals of the thermochemical cycle is that the waste heat from a nuclear reactor can be used to supply the necessary hot utility, as is done with an alternative thermochemical cycle (ATC) [7], and purified oxygen, a much sought after commodity, is produced as a by-product. This paper specifically deals with ATCs, in which the cycle contains no sulfur or sulfur products. A temperature limit of 1123K is imposed so that nuclear waste heat can provide the hot utility requirements. The ATCs under investigation specifically do not consist of more than five reactions, are chemically and thermodynamically viable, and do not contain noble metals or other substances with low abundances [7]. To abide by RICA standards the ATCs do not contain mercury, selenium, or cadmium.

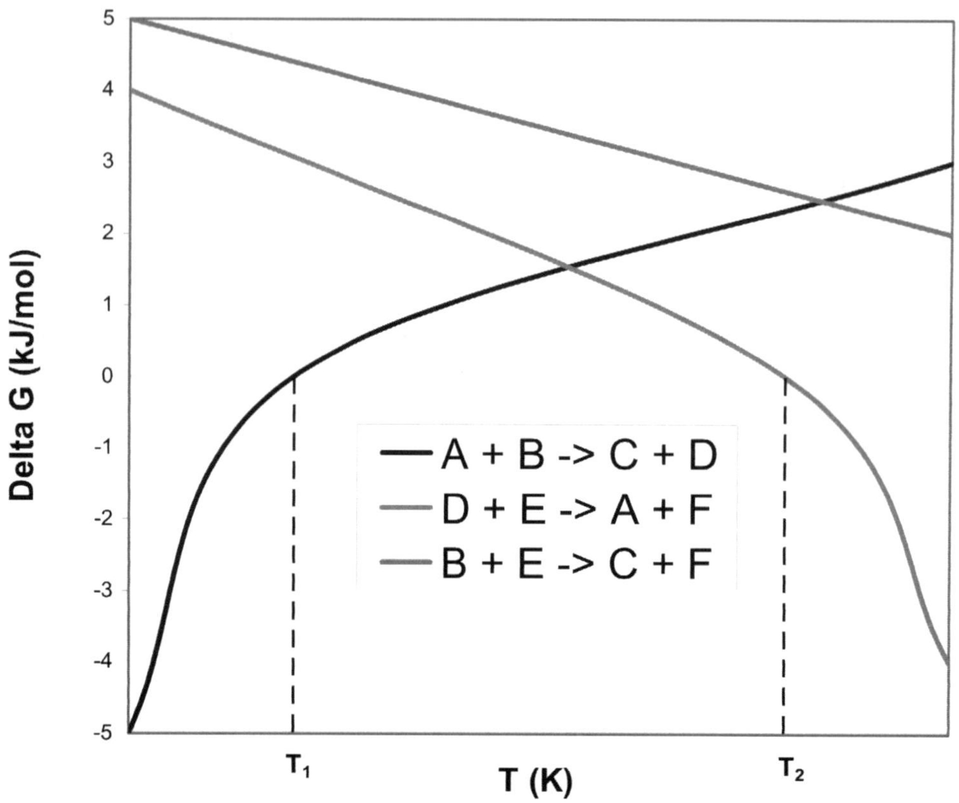

Figure 2. Plot of ΔG as a function of reaction temperature for three hypothetical reactions, where the third is the sum of the first two, which demonstrates the need for thermochemical cycles

The difficulty with thermochemical cycles is with the selection and evaluation of the reaction clusters. Traditional assessment tools have primarily included literature searches [8], inspection, and evaluation by comparison [9] to screen possible clusters. Few attempts have been made to systematically identify, screen, and evaluate hydrogen producing thermochemical cycles. A proper study should be thorough, exhausting all possible thermodynamically feasible options.

In this work a systems engineering approach is taken to develop a systematic methodology for the initial evaluation of ATCs for hydrogen production, as shown by Figure 3. This proposal consists of five major components, (i) conceptualization, (ii) reaction cluster synthesis, (iii) flowsheet design, simulation, and analysis, (iv) process integration and (v) performance evaluation. A formal definition of thermodynamic feasibility is given, followed by a mathematical construct for generating feasible clusters. Steps are then prescribed for design, optimization, and heat integration, along with a formal definition for efficiency and prescribed minimum threshold efficiency. After the strategy is outlined, its strength and flexibility is then demonstrated through a case study involving the Fe-Cl thermochemical cycle.

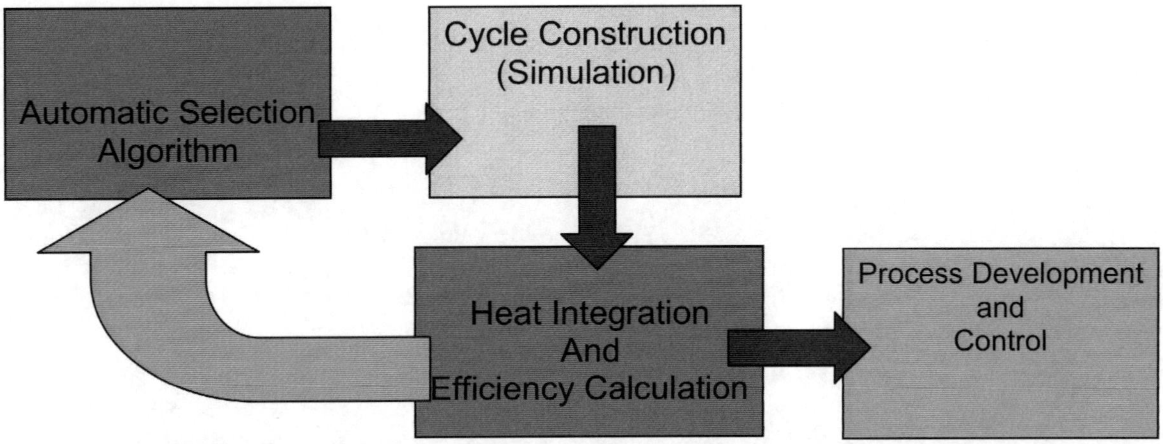

Figure 3. Flowchart of the systematic methodology for identifying, screening, and evaluating thermochemical cycles, in which the participating species are chosen for a given ATC and are supplied to an automatic selection algorithm, each identified cycle is constructed using simulation software, then a heat integration is executed and the efficiency of each cycle is determined, if a promising ATC is identified (one whose efficiency exceeds a threshold value) further research and development are conducted on that cycle, otherwise parameters in the algorithm are modified or new molecular and/or atomic species are considered for cycle construction

2. Computational Methodology

The space containing the reactions involved in a desired cycle is defined as follows, which is adapted from the work of Holiastos and Manousiouthakis [10] and Stephanopoulos ([11], [12]).

We begin by defining our reaction cluster as follows, where N^R is the number of reactions, N^S is the number of chemical species, γ_{ij} is the reactant coefficient for species Λ_i in reaction j, and δ_{ij} is the product coefficient for species Λ_i in reaction j. The molecular species Λ_i involved are contingent on the N^A desired atomic species, which satisfy criteria of safety, abundance, and affordability in accordance with an ATC [7]. The overall desired reaction has coefficients of ν_i ($\nu_i > 0$ is a product) which the cluster must be constrained to.

To achieve a desired cycle, all of the involved reactions must be thermodynamically feasible. What follows is a derivation of thermodynamic feasibility along with the mathematical constraints for a reaction cluster.

$$\sum_{i=1}^{N^S} \gamma_{ij}\Lambda_i \to \sum_{i=1}^{N^S} \delta_{ij}\Lambda_i \qquad j \in \{1,2,...,\mathrm{N}^\mathrm{R}\} \qquad (5)$$

$$\sum_{j=1}^{N^R} \left(\delta_{ij} - \gamma_{ij}\right) = v_i \qquad i \in \{1,2,...,N^S\} \qquad (6)$$

2.1 Thermodynamic Modeling of Reaction Clusters

A reaction cluster is often evaluated based on prescribed criteria, such as process safety, environmental protection and economic viability. Process safety tends to constrain the reactants and by-products in the reaction cluster to be nontoxic and nonflammable. Environmental considerations require reducing the byproducts and eliminating hazardous compounds. Economic viability is significantly determined by the reaction yield.

Thermodynamics plays a central role in reaction yield since it sets an upper bound on a reaction attainable yield. If a process is thermodynamically infeasible, process safety and economic viability are greatly impaired. Thermodynamics can limit the process yield which is directly correlated with product yield and economic viability. With a lower yield larger flowrates and more energy will have to be used, reducing profit. Process safety and environmental concerns are also affected by unexpected, thermodynamically favored, byproducts, which could have adverse health effects. Larger operations, resulting from unfavorable reaction dynamics, also pose a threat to process safety.

When the yield of the process, specified by thermodynamics, is greater than or equal to the yield ensuring economic feasibility, then the corresponding reaction is called thermodynamically feasible. The purpose of the work is to screen out some hydrogen producing reaction cluster that is thermodynamically feasible within a certain temperature range within a database of some selected chemical species, while meeting imposed environmental and safety criteria. The first effort in generating reaction clusters is to establish acceptable criteria for thermodynamic feasibility.

Consider a closed, reacting system, and the overall differential for internal energy (U) and its relationship to other state variables.

$$d(nU) = Td(nS) - Pd(nV) + \sum_i \mu_i dn_i \qquad (7)$$

Re-arranging (7), using the first and second laws of thermodynamics, we determine the total change in chemical potential to be negative, where μ_i is the chemical potential of species i and dn_i is the transfer rate of i to and from the system.

$$-\sum_i \mu_i dn_i = Td(nS) - d(nU) - Pd(nV) = Td(nS) - d(nQ) \geq 0 \qquad (8)$$

Given the definition of the differential of Gibbs energy (G):

$$d(nG) = -nSdT + nVdP + \sum_i \mu_i dn_i \qquad (9)$$

For an isothermal, isobaric closed reacting system, by combining (8) and (9), we find that thermodynamics constrains changes in Gibbs energy to be negative.

$$d(nG) = \sum_i \mu_i dn_i \leq 0 \qquad (10)$$

The equilibrium constant, $K(T)$ is then specified in terms of the change in Gibbs energy for a reaction ($\Delta G^{\circ}(T)$), which specifies the reaction to occur favorably with some driving force e^z.

$$K(T) \equiv e^{-\Delta G^{\circ}(T)/RT} \geq e^z \qquad z > 0 \qquad (11)$$

By assuming the equilibrium constant to be of exponential order, we know the function $K(T)$ is bounded to finite values over any temperature range [13].
Combining that with (10) gives the following:

$$\frac{-\Delta G^{\circ}(T)}{RT} - z \geq 0 \qquad (12)$$

Re-arranging (12) gives us our definition for thermodynamic feasibility:

$$\Delta G^{\circ}(T) + zRT \leq 0 \qquad (13)$$

The requirement of thermodynamic feasibility for each reaction in cluster synthesis is as follows, where T_j is the temperature of reaction j.

$$\sum_{i=1}^{N^S} \left(G_{iT_j}\right)\left(\delta_{ij} - \gamma_{ij}\right) \leq 0 \qquad j \in \{1, 2, ..., N^R\} \qquad (14)$$

Our formulation employs a database of ΔG° values for each molecular species within a temperatures range, T^L to T^U. This differs from the literature, where the Gibbs energy is usually assumed to be a linear function of temperature, forcing the reactions to occur at the bounds of T^L and T^U [10]. We have found the Gibbs energy of formation to be highly nonlinear, due primarily to phase changes, so by specifying Gibbs at several temperatures and assuming that the reactions can take place at any of those temperature values within the range of T^L to T^U, we can overcome the nonlinearities.

In addition to the cluster satisfying thermodynamic feasibility (14) and the overall desired reaction (6), an atomic species balance must be satisfied, where B_{ki} is the number of atoms of k in species i per molecule.

$$\sum_{i=1}^{N^S} B_{ki}\left(\delta_{ij}-\gamma_{ij}\right)=0 \qquad k\in\{1,2,\ldots,N^A\}\; j\in\{1,2,\ldots,N^R\} \qquad (15)$$

Limits on the allowable reaction coefficients must now be placed.

$$\delta_j^l \le \sum_{i=1}^{N^S}\delta_{ij} \le \delta_j^u \qquad j\in\{1,2,\ldots,N^R\} \qquad (16)$$

$$\gamma_j^l \le \sum_{i=1}^{N^S}\gamma_{ij} \le \gamma_j^u \qquad j\in\{1,2,\ldots,N^R\} \qquad (17)$$

$$\delta_{ij}\in\left\{\phi_1,\ldots,\phi_n\right\} \qquad i\in\{1,2,\ldots,N^S\}\; j\in\{1,2,\ldots,N^R\} \qquad (18)$$

$$\gamma_{ij}\in\left\{\varphi_1,\ldots,\varphi_m\right\} \qquad i\in\{1,2,\ldots,N^S\}\; j\in\{1,2,\ldots,N^R\} \qquad (19)$$

$$0\le\delta_{ij}\le\delta_{ij}^u \qquad i\in\{1,2,\ldots,N^S\}\; j\in\{1,2,\ldots,N^R\} \qquad (20)$$

$$0\le\gamma_{ij}\le\gamma_{ij}^u \qquad i\in\{1,2,\ldots N^S\}\; j\in\{1,2,\ldots,N^R\} \qquad (21)$$

Inequalities (16) and (17) limit the overall sum of product and reactant coefficients, in each reaction, to lie between upper and lower bounds. Equations (18) and (19) constrain δ_{ij} and γ_{ij} to be integer values, while (20) and (21) put bounds on the allowable integer values.

Next we define flag variables, β_{ij} and θ_{ij}, that indicate whether species i is a product or a reactant in reaction j, respectively. These variables eliminate redundancy within a particular solution. With these new variables we rewrite our coefficient constraints, so that if a species i is involved as a product in reaction j then $\beta_{ij}=1$ and if it is a reactant in reaction j then $\theta_{ij}=1$.

$$\beta_{ij} \in \{0,1\} \qquad i \in \{1,2,...N^S\} \; j \in \{1,2,...,N^R\} \qquad (22)$$

$$\theta_{ij} \in \{0,1\} \qquad i \in \{1,2,...N^S\} \; j \in \{1,2,...,N^R\} \qquad (23)$$

$$0 \le \delta_{ij} \le \delta_{ij}^u \beta_{ij} \qquad i \in \{1,2,...N^S\} \; j \in \{1,2,...,N^R\} \qquad (24)$$

$$0 \le \gamma_{ij} \le \gamma_{ij}^u \theta_{ij} \qquad i \in \{1,2,...N^S\} \; j \in \{1,2,...,N^R\} \qquad (25)$$

To ensure that a species does not occur as both a product and a reactant in the same reaction we add the following, which is not evident in [10].

$$\beta_{ij} + \theta_{ij} \le 1 \qquad i \in \{1,2,...N^S\} \; j \in \{1,2,...,N^R\} \qquad (26)$$

Limits on the total number of species involved in a reaction are then invoked, and the range of possible reaction temperatures is enforced.

$$\sum_{i=1}^{N^S} \beta_{ij} \le s_P \qquad j \in \{1,2,...,N^R\} \qquad (27)$$

$$\sum_{i=1}^{N^S} \theta_{ij} \le s_R \qquad j \in \{1,2,...,N^R\} \qquad (28)$$

$$T_j \in \{\psi_1,...,\psi_n\} \qquad j \in \{1,2,...,N^R\} \qquad (29)$$

$$T^L \le T_j \le T^U \qquad j \in \{1,2,...,N^R\} \qquad (30)$$

2.2 ILP Formulation

The above model is the foundation for a structured definition of a feasible thermochemical cycle. To screen and identify all feasible clusters, the algorithm is formulated as the following minimization problem:

$$(P1) \qquad \min \sum_{i=1}^{N^S} \sum_{j=1}^{N^R} \left(d_i \delta_{ij}^2 + c_i \gamma_{ij}^2 \right)$$

subject to:

(6), (14) – (30)

The objective function presented here, featuring adjustable weights on the reaction coefficients, is used only as a tool for identifying the clusters. Constraining the objective to be greater than a known minimum allows for additional cycles to be generated. $P1$ is an integer linear program (ILP), because each constraint, including the objective function, is linear, and the decision variables can only assume integer values. The reaction coefficients are confined to integer values, so as to eliminate unrealizable and obtuse solutions. The reaction temperatures are also integer values as we have data for Gibbs energy at specific integer temperature values.

With this flexible definition one can search for both the reactions and their temperature, or the reaction temperatures can be specified as a parameter, leaving only the reaction coefficients as the variables. With the conceptualization in place, we now put forth the necessary tools for evaluating generated ATCs.

2.3 Cycle Evaluation

A systematic method must also be employed in the evaluation of the cycles. This includes reaction cluster construction, process integration, and an efficiency calculation. This section describes such a process for the methodic evaluation of generated ATCs.

For the purpose of initial ATC screening, it is assumed that each reaction in a generated cycle goes to completion within the thermodynamically feasible temperature range and that there are no competing products. The analysis assumes that the entrance and exit streams to the system are at a temperature of 298K and a pressure of 1 bar. Each reactor is sequenced appropriately in a simulation software, with heat exchangers and reactors in between the key reactions to model temperature and phase transitions, respectively. The details of flow separation and pumping are not considered in any stage of the presented analysis.

Once the cycle has been constructed a heat integration is performed and the cluster's efficiency is determined. Taking the latent, sensible, and reaction heats from the simulated flowsheet, a pinch analysis is implemented to aid in process integration. Temperature intervals, which break at phase transitions and reaction temperatures, are determined for the hot (i.e. require cooling) and cold (i.e. require heating) process streams. Overall heating and cooling requirements are determined for each temperature interval and then the two streams are plotted against each other, as temperature versus heat duty. The two curves can be moved horizontally in enthalpy so that the hot curve lies above the cold curve. Maximum overlap should be ensured while satisfying a minimum approach temperature or pinch. This will give the designer the areas where heat can be integrated between processes.

The efficiency (η), defined below, is prescribed by Argonne National Laboratory (ANL) [7] and considers the cycle heat duty (Q) and required work (W) against the standard heat of formation of water $(\Delta H^{\circ}_{298K}(H_2O))$. The heat duty, determined from the pinch analysis, is the summation of the needed hot and cold utility requirements (i.e. the areas where the curves do not overlap). The definition of work can include separation, chemical, electrochemical, mechanical, and electrical requirements [7].

$$\eta = \frac{-\Delta H^{\circ}_{298K}(H_2O)}{Q + \dfrac{W}{0.5}} \qquad (31)$$

The primary work to be considered is the work of separation which is equal to ΔG_{sep}, the resulting change in Gibbs energy.

$$\Delta G_{sep} = -RT\sum_{i} n_i \ln y_i \qquad (32)$$

This simplistic definition, provided by ANL [7], allows for the calculation of ΔG_{sep} (which is the required work for an ATC that does not involve electrochemical steps, for a base efficiency calculation), where R is the gas constant, T is the absolute temperature, n is the molar flow of each component, and y is the mole fraction. Equation (31) assumes an efficiency of 50% for the thermal equivalent of the required work, which we feel to be a conservative estimate. In the initial, ideal analysis, only the separation costs for obtaining hydrogen and oxygen are included in the work calculation. A promising cycle has a base efficiency that exceeds the threshold value of 35% $(\eta > 0.35)$ [7].

Using this analysis, to identify, screen, and evaluate hydrogen producing ATCs, we have the tools to locate a cluster that is competitive with current technologies. A case study on the iron-chlorine cycle is now presented, which demonstrates the strength and flexibility of the proposed methodology.

3. Results and Discussion: A Case Study on the Fe-Cl Thermochemical Cycle

We now apply the proposed methodology to a case study to assess the performance of iron-chlorine based ATCs for hydrogen production. The atomic species involved in the cycle are Fe, Cl, H, and O, which are ideal elements in terms of economic and environmental concerns. From these species we develop a list of 14 molecular species to be included in the cluster synthesis:

$FeCl_2$, $FeCl_3$, Fe_3O_4, H_2O, HCl, Cl_2, H_2, O_2, Fe, FeO, Fe_2O_3, $Fe(OH)_2$, $Fe(OH)_3$, Fe_2Cl_6

The algorithm $P1$, which is an ILP, is solved using an AMPL based solver, Cbc, available through the NEOS Server at Argonne National Laboratories. Given thermophysical data from Aspen Properties for the above species, at 100 degree temperature intervals between 400K and 1200K, 10 cycles are identified using the parameter set shown in Table 1. The value of zRT is chosen as to allow for the largest possible set of cycles while avoiding null reactions.

This set of cycles is generated by assuming that at least one reaction occurs at 400K and that at least one reaction occurs at 1200K, which allows for better heat exchange. Additional cycles can be generated using the same parameters by constraining the objective function to be greater than the observed minimum.

Table 1. List of parameters for $P1$

Parameter	Value
N^A	4
N^S	14
N^R	4
d_i	1
c_i	1
zRT_j	100 J/mol
δ_j^l	1
δ_j^u	20
γ_j^l	1
γ_j^u	20
δ_{ij}^u	16
γ_{ij}^u	16
$\{\phi_1,...,\phi_n\}$	$\{1,2,...,16\}$
$\{\varphi_1,...,\varphi_m\}$	$\{1,2,...,16\}$
s_P	3
s_R	2
T^L	400 K
T^U	1200 K
$\{\psi_1,...,\psi_n\}$	$\{400,500,...,1200\}K$

This set of cycles is generated by assuming that at least one reaction occurs at 400K and that at least one reaction occurs at 1200K, which allows for better heat exchange. Additional cycles can be generated using the same parameters by constraining the objective function to be greater than the observed minimum.

Each cluster is evaluated by the methodology outlined above, using a pinch of 20 K in the heat integration, and their efficiencies are given in Table 2. A cycle, found in the literature as one of the most promising Fe-Cl cycles [14], is generated, in addition to the other ten cycles, with an objective value of 81, using the same parameter set and letting $T_1 = 1200K$, $T_2 = 500K$, $T_3 = 700K$, and $T_4 = 1000K$. This cycle, shown below with literature suggested temperature intervals [14] and represented by Figure 4, has a base efficiency of 35.6%, which is comparable with the literature.

$$6FeCl_{2(s,l)} + 8H_2O_{(g)} \rightarrow 2Fe_3O_{4(s)} + 12HCl_{(g)} + 2H_{2(g)} \qquad (800K - 1200K)$$

$$2Fe_3O_{4(s)}(s) + 16HCl_{(g)} \rightarrow 4FeCl_{3(s,g)} + 2FeCl_{2(s)} + 8H_2O_{(g)} \qquad (400K - 600K)$$

$$4FeCl_{3(s,g)} \rightarrow 4FeCl_{2(s)} + 2Cl_{2(g)} \qquad (500K - 700K)$$

$$2Cl_{2(g)} + 2H_2O_{(g)} \rightarrow 4HCl_{(g)} + O_{2(g)} \qquad (900K - 1200K)$$

This study demonstrates the power of our methodology to identify pre-existing ATCs and unknown cycles. The methodology can also be modified to generate and evaluate hybrid ATCs [15], or thermochemical cycles with electrochemical steps. This flexible systems engineering approach will no doubt result in the discovery of economically feasible ATCs for the production of hydrogen. Once a promising cycle is identified then further analysis will be conducted, including detailed parametric studies on reaction thermodynamics and kinetics, along with the design of reactors and separation processes to give a full energy and economic evaluation of the proposed cycle.

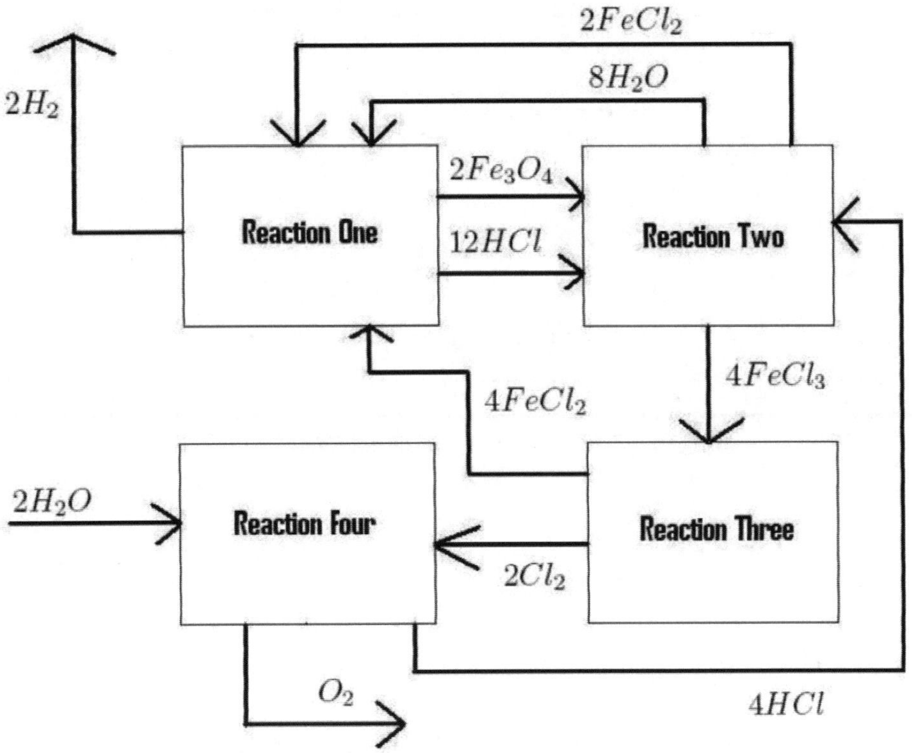

Figure 4. Diagram of the identified Fe-Cl ATC for hydrogen production, that can be found in the literature [14]

Table 2. Base efficiencies of the 10 identified Fe-Cl ATCs using $P1$ with the parameter set from Table 1, and assuming that at least one reaction occurs at the upper temperature limit and that at least one reaction occurs at the lower temperature limit

Cycle Number	Reactions	Base Efficiency (%)
1	$6H_2O + 10FeO \rightarrow 2Fe_3O_4 + 2H_2 + 4Fe(OH)_2$	42.5
	$2Fe_3O_4 + 3Fe(OH)_2 \rightarrow 3H_2O + O_2 + 9FeO$	
	$Fe(OH)_2 \rightarrow H_2O + FeO$	
2	$8H_2O + 4Fe \rightarrow 4H_2 + 4Fe(OH)_2$	44.1
	$2H_2 + 4Fe(OH)_2 \rightarrow 6H_2O + O_2 + 4Fe$	
3	$8H_2O + 4Fe \rightarrow 5H_2 + 2FeO + 2Fe(OH)_3$	48.8
	$H_2 + 2Fe(OH)_3 \rightarrow 4H_2O + O_2 + 2Fe$	
	$H_2O + 2FeO \rightarrow H_2 + Fe_2O_3$	
	$3H_2 + Fe_2O_3 \rightarrow 3H_2O + 2Fe$	
4	$2H_2O + Fe \rightarrow H_2 + Fe(OH)_2$	*Inordinate*[2]
	$Fe + 3Fe(OH)_2 \rightarrow 3H_2 + O_2 + 4FeO$	
	$3H_2O + 5FeO \rightarrow H_2 + 2Fe(OH)_2 + Fe_3O_4$	
	$Fe_3O_4 + 3H_2 \rightarrow 3H_2O + 2Fe + FeO$	

5	$3H_2O + 5FeO \rightarrow Fe_3O_4 + H_2 + 2Fe(OH)_2$	35.0
	$2H_2 + 4Fe(OH)_2 \rightarrow 6H_2O + O_2 + 4Fe$	
	$5H_2O + 3Fe \rightarrow 3H_2 + FeO + 2Fe(OH)_2$	
	$Fe_3O_4 + Fe \rightarrow 4FeO$	
6	$9H_2O + 4Fe \rightarrow 6H_2 + Fe_2O_3 + 2Fe(OH)_3$	*Inordinate*[2]
	$Fe_3O_4 + 3H_2 \rightarrow 3H_2O + 2Fe + FeO$	
	$FeO + Fe_2O_3 \rightarrow Fe_3O_4$	
	$H_2 + 2Fe(OH)_3 \rightarrow O_2 + 2Fe + 4H_2O$	
7	$6H_2O + 10FeO \rightarrow 2Fe_3O_4 + 2H_2 + 4Fe(OH)_2$	43.0
	$2Fe_2O_3 + 3Fe(OH)_2 \rightarrow 3H_2O + O_2 + 7FeO$	
	$Fe_3O_4 + Fe(OH)_2 \rightarrow H_2 + 2Fe_2O_3$	
	$Fe_3O_4 + H_2 \rightarrow H_2O + 3FeO$	
8	$6H_2O + 10FeO \rightarrow 2Fe_3O_4 + 2H_2 + 4Fe(OH)_2$	41.7
	$2Fe_3O_4 + 3Fe(OH)_2 \rightarrow 3H_2O + O_2 + 9FeO$	
	$FeO + Fe(OH)_2 \rightarrow H_2 + Fe_2O_3$	
	$H_2 + Fe_2O_3 \rightarrow H_2O + 2FeO$	

9	$5H_2O + 9FeO \rightarrow 2Fe_3O_4 + 2H_2 + 3Fe(OH)_2$	*Inordinate*[2]
	$Fe + 3Fe(OH)_2 \rightarrow 3H_2 + O_2 + 4FeO$	
	$Fe_3O_4 + H_2 \rightarrow H_2O + 3FeO$	
	$Fe_3O_4 + 2H_2 \rightarrow 2H_2O + Fe + 2FeO$	
10	$8H_2O + 4Fe \rightarrow 4H_2 + 4Fe(OH)_2$	*Inordinate*[2]
	$4Fe(OH)_2 \rightarrow 4H_2 + O_2 + 2Fe_2O_3$	
	$6H_2 + 2Fe_2O_3 \rightarrow 6H_2O + 4Fe$	

[2] For $\Delta G < 0$, $T_j > 1123K$

4. Conclusions

In this paper we outline a methodology that can be used to identify, screen, and evaluate hydrogen producing alternative thermochemical cycles. The process is five fold, giving rigorous instructions for (i) conceptualization, (ii) reaction cluster synthesis, (iii) flowsheet design, simulation, and analysis, (iv) process integration and (v) performance evaluation. Starting with a formal definition of thermodynamic feasibility , an automatic selection algorithm is proposed, followed with a base level of analysis using simulation software, the use of a pinch analysis for heat integration, and ending with a standard definition for an efficiency calculation.

The promise of the process is then demonstrated through a case study involving the Fe-Cl thermochemical cycle. We are then able to identify a, with our algorithm, a cycle that can be found throughout the literature, and determine an efficiency for that cycle that is comparable with the literature. This method allows for a standard and efficient procedure for the future evaluation of alternative thermochemical cycles for hydrogen production and can be extended to hybrid cycles and to higher order evaluations.

Literature Cited:

[1] Marshall SL, Blencoe JG. Equilibrium analysis of thermochemical cycles for hydrogen production. Separation Science and Technology 2005;40(1-3):483-505.

[2] Berry GD, Aceves SM. The Case for Hydrogen in a Carbon Constrained World. Journal of Energy Resources Technology 2005;127(2):89-94.

[3] Hake J-F, Linssen J,Walbeck M. Prospects for hydrogen in the German energy system. Energy policy 2006;34:1271-1283.

[4] Ersoz A, Olgun H, Ozdogan S. Reforming options for hydrogen production from fossil fuels for PEM fuel cells. Journal of Power Sources 2006;154(1):67-73.

[5] Pinto F, Franco C, Lopes H, Andre R, Gulyurtlu I, Cabrita I. Effect of used edible oils in coal fluidised bed gasification. Fuel 2005;84(17), The 5th European Conference on Coal Research and its Applications:2236-2247.

[6] Gaykara SZ. Experimental solar water thermolysis. Int. J. Hydrogen Energy 2005;29(14):1459-1469.

[7] Lewis MA. An Assessment of the Potential of Alternative Cycles. Prepared for U.S. Department of Energy Office of Nuclear Energy, Science and Technology 2006.

[8] Brown LC, Besenbruch GE, Schultz KR, Showalter SK, Marshall AC, Pickard PS, Funk JF. High Efficiency Generation of Hydrogen Fuels using Thermochemical Cycles and Nuclear Power. Prepared for Presentation at the AIChE 2002 Spring National Meeting, New Orleans, LA, March 11-15, 2002.

[9] Shinnar R, Shapira D, Zakal S. Thermochemical and Hybrid Cycles for Hydrogen Production: A Differential Economic Comparison with Electrolysis. Industrial and Engineering Chemistry Process Design and Development 1981;20:581-593.

[10] Holiastos K, Manousiouthakis V. Automatic Synthesis of Thermodynamically Feasible Reaction Clusters. AIChE Journal 1998;44(1):164-173.

[11] Mavrovouniotis M, Stephanopoulos G. Synthesis of Reaction Mechanisms Consisting of Reversible and Irreversible Steps. 1. A Synthesis Algorithm. Ind. Eng. Chem. Res. 1992;31(7):1625-1637.

[12] Fornari T, Stephanopoulos G. Synthesis of Chemical Reaction Paths: Economic and Specification Constraints. Chem. Eng. Commun. 1994;129:159-182.

[13] Greenberg MD. Advanced Engineering Mathematics, Second Edition. Upper Saddle River, New Jersey: Prentice Hall, 1998.

[14] Gahimer J, Mazumder M, Pangborn J. Experimental Demonstration of an Iron Chloride Thermochemical Cycle for Hydrogen Production. Eleventh Intersociety Energy Conversion Engineering Conference 1976;1:933-939.

[15] Uehara I, Ishii E, Ishikawa H, Kito A, Nakane M. Search for hybrid electrochemical-thermochemical cycles for hydrogen production from water. Osaka Kogyo Gijutsu Shikensho Kiho 1981;32(4):262-8.

Status of the Development Effort for the Thermochemical Cu-Cl Cycle

Magali Ferrandon[1], Michele Lewis[1], David Tatterson[2], and Alan Zdunek[3]

(1) Chemical Sciences and Engineering Division, Argonne National Laboratory, 9700 South Cass Avenue, Argonne, IL 60439.
(2) Orion Consulting Co, 1240 Gilbert Avenue, Downers Grove, IL 60515.
(3) Center for Electrochemical Science and Engineering Chemical and Biological Engineering Department Illinois Institute of Technology Suite 127 Perlstein Hall 10 West 33rd Street Chicago, Illinois 60616.

1. Introduction

The hybrid copper-chlorine (Cu-Cl) cycle is one of the most promising thermochemical cycles for hydrogen production using nuclear or solar heat. The advantage of the hybrid CuCl cycle relative to other cycles is the relatively lower temperature heat (550°C) source required which should mitigate some of the demands on the materials of construction. Several types of nuclear reactors can be used as a heat source, such as the supercritical water reactor being developed in Canada, CANDU Mark 2, the lead cooled reactor, or the high temperature gas reactor. Solar heat can be provided using the commercially proven tower technology. All these provide heat near or above 600°C, the maximum temperature required for the cycle.

The copper-chlorine cycle consists of the three major reactions shown in Table 1. The electrolysis reaction *(1)* in which cupric chloride ($CuCl_2$) is produced at the anode and H_2 is released at the cathode. The $CuCl_2$(a) from *(1)* is hydrolyzed to copper oxychloride (Cu_2OCl_2) according to the hydrolysis reaction *(2)*. Molten cuprous chloride (CuCl) is then produced from the decomposition reaction *(3)*.

Table 1. Reactions in the Copper-Chlorine Cycle

Reaction	Conditions
(1) $2CuCl(a) + 2HCl(a) + 2H_2O \rightarrow CuCl_2{\cdot}2H_2O(a) + H_2(g)$	100°C, 24 bar
(2) $2CuCl_2(s) + H_2O(g) \rightarrow Cu_2OCl_2(s) + 2HCl(g)$	400°C, 1 bar
(3) $Cu_2OCl_2(s) \rightarrow \frac{1}{2} O_2(g) + 2CuCl(s)$	540°C, 1 bar

All reactions have been experimentally demonstrated. The two thermal reactions, the hydrolysis of $CuCl_2$ *(2)*, and the decomposition of Cu_2OCl_2 *(3)* have been proven at ANL. In bench scale experiments, all of the oxygen was recovered at 530°C from reaction *(3)*. The early experiments indicated technical challenges in the hydrolysis *(2)* and electrolysis reactions *(1)*. The electrolytic reaction *(1)* was demonstrated at the Atomic Energy of Canada, Ltd. (AECL) at Chalk River recently. Meeting performance of 500 mA/cm^2 at 0.7V is the primary challenge for the electrolysis reactor.

We consider the hydrolysis reaction to be the most challenging reaction because of two factors: (i) a competing reaction of $CuCl_2$ and (ii). the need for excess water The competing reaction is the thermal decomposition of $CuCl_2$:

$$2CuCl_2\ (s) \Leftrightarrow 2CuCl\ (s) + Cl_2\ (g). \qquad (4)$$

Because CuCl is a product of the subsequent reaction, this competing reaction is not a showstopper, provided the chlorine can be scavenged and the amount of chlorine formed is minimal. We also believe that this competing reaction can be minimized by the choice of operating conditions and the reactor design. A sensitivity study and the experimental results indicate that the steam must be in excess for high yields of the desired Cu_2OCl_2 and HCl. The excess steam increases capital costs significantly because of the larger number of vessels required and the high energy usage for vaporizing the water.

The paper is divided into two sections. The first section describes the current conceptual design. In order to determine the potential of the Cu-Cl cycle, an Aspen Plus™ flowsheet was developed using this process design and the cycle's efficiency was calculated. The energy and mass balances, the heat exchanger duties and shaft work were calculated, and heat recovery was optimized with pinch analysis. The hydrogen production cost was estimated using the hydrogen analysis (H2A) methodology [1]. Capital costs for the thermal processes were estimated using Capcost software [2]. Operating costs (thermal and electrical energy, cooling water, etc.) were set to be consistent with the H2A analysis.

In the second section of this paper, experimental operating parameters were varied to see which had the greatest impact on the Cu_2OCl_2 formation during the hydrolysis reaction. The goal is to inhibit $CuCl_2$ thermal decomposition while minimizing the steam-to-copper molar ratio (S/Cu). The latter is critical for high efficiency since vaporizing the water is energy intensive. The reactor used is a engineering-scale continuous spray reactor, similar to the industrial spray roasters.

2. Experimental

Past work has shown that the previous fixed bed reactor design was not optimal [3]. Inefficiencies in heat and mass transfer inhibited the desired reaction. Unreacted $CuCl_2$ was observed and this material tended to be agglomerated in the middle of the fixed bed.

A spray reactor was used for the new series of tests, as shown in Figure 1. In this concept, a solution of $CuCl_2$ is injected into the reactor using one of two types of atomizers. The first one is a pneumatic quartz nebulizer used for Inductively Coupled Plasma-Mass Spectrometry (ICP-MS) analyses (Glass Expansion) and the other one is an ultrasonic nozzle (Sono-Tek). The pneumatic one requires Ar as a sweep gas to create a fine mist of $CuCl_2$ solution while the ultrasonic nozzle uses ultrasonic vibrations to atomize the solution. When the mist enters the heated zone, mass transfer is enhanced relative to the fixed bed because the droplets are dispersed and very small. Superheated steam with or without Ar was used to improve heat transfer. Steam can be added from the top (co-current flow design) or at the bottom (counter-current flow design). Solid products are collected at the bottom of the reactor where the temperature is kept at 150°C for quenching the reaction while keeping the material dry.

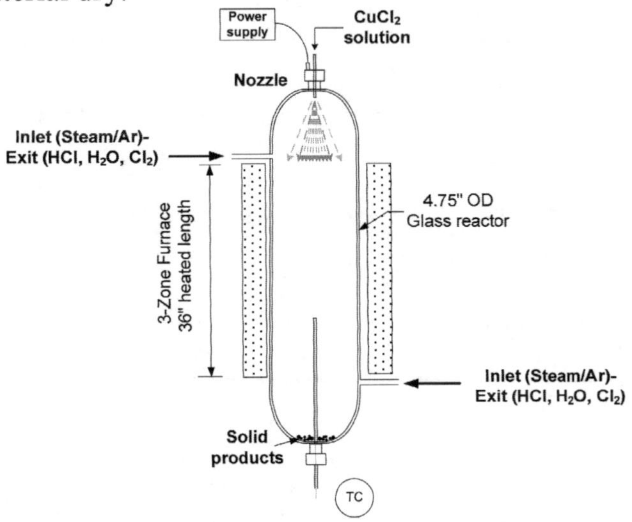

Figure 1. Schematic of the nebulizer reactor for the Cu-Cl hydrolysis experiments. TC=thermocouple.

This design is a continuous reactor, similar to that in the conceptual process design (described below) except for the presence of Ar and the pressure/temperature (1 atm/room temperature vs 24 bar/400°C) at which $CuCl_2$ stream enters the reactor.

3. Results

3.1. Conceptual Process

Aspen Plus [4] was used to develop mass and energy balances for a process based on these 3 reactions. Little is known about the electrolysis reaction and in the calculations, we have assumed the electrolyzer operates at 0.7 V and 500 mA/cm^2. The Aspen simulation provided stream flows and properties as well as heat exchanger duties and work requirements for pumps. The overall process flow sheet is shown in Figure 2.

A CuCl cycle plant producing 125 MT H_2/day requires 191 MW of thermal energy and 100.5 MW of electrical energy. The energy efficiency of the process is defined as energy out divided by energy in. Based on the low heating value for hydrogen, the efficiency of this process is:

$$\text{Efficiency} = \frac{\text{Mol. of } H_2 \text{ Produced} * \text{LHV}}{(\text{Shaft work} + \text{Electrochemical work} + \text{Pinch Heat})}$$

$$\text{Efficiency} = 125{,}000 \times 33.3/ (24(3000/0.4{+}97{,}500/0.4 {+}191{,}000) = 39\% \text{ (LHV)}$$

In this equation it is assumed that the efficiency of converting thermal energy into electrical power is 0.4.

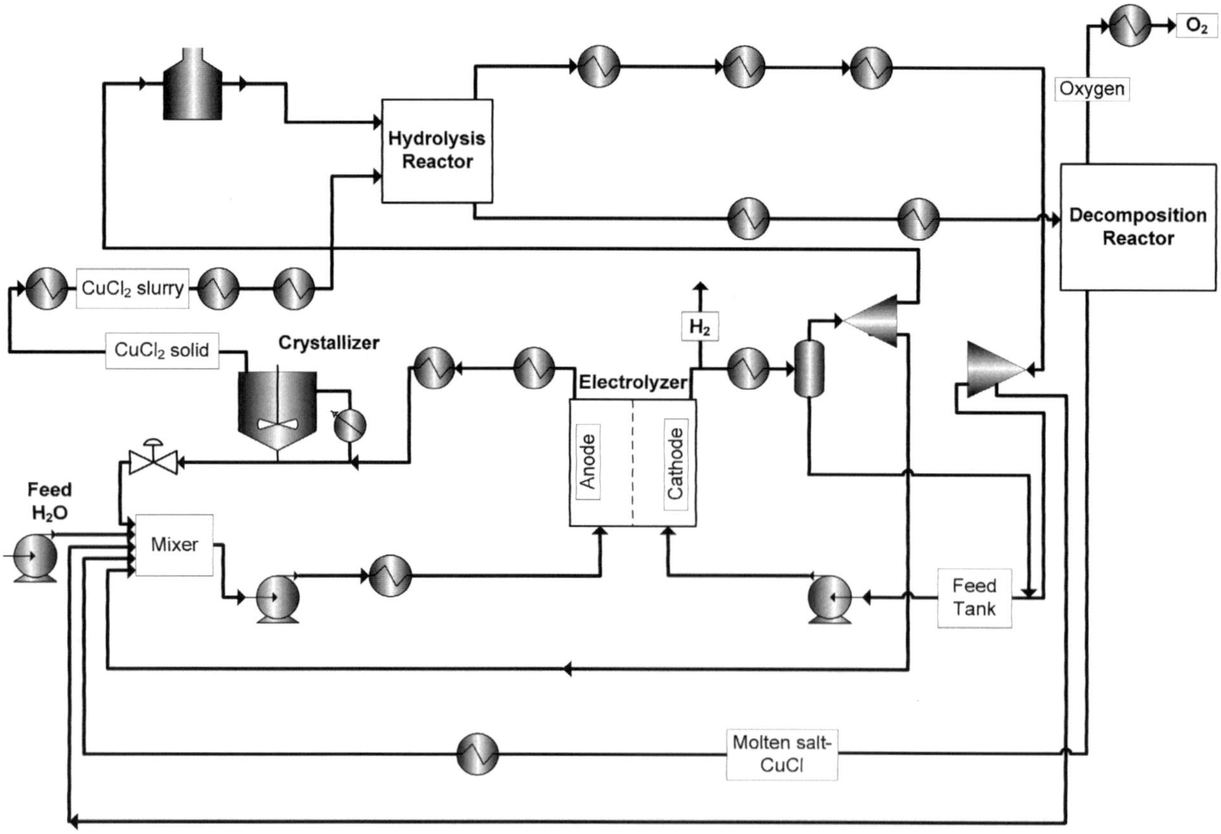

Figure 2. Aspen flow sheet of the CuCl cycle.

The conceptual process design for the CuCl cycle is divided into 4 sections:

1. The electrolyzer
2. The crystallizer
3. The hydrolysis/oxychloride decomposition reactors
4. The CuCl heat recovery system

We have assumed that the electrolyzer will operate near 100°C at 24 bar in order to produce hydrogen at 300 psi. A conceptual design for the electrolyzer has been developed and is discussed below. Many of the process steps are similar to commercially practiced technologies. These technologies are pointed out where appropriated.

Electrolyzer /Crystallizer

Anode and Crystallizer Section: Feed water, recycled granulated CuCl, any make up HCl and crystallizer recycle is added to the anode feed tank. The solution containing dissolved CuCl, HCl and residual $CuCl_2$ is then pressurized to 24 bar and transferred to the anode section of the electrolyzer. Chloride ion migrates from the cathode across the electrolyzer membrane and reacts at the anode with the CuCl to form $CuCl_2$.

The anolyte solution is then cooled to 55°C in the crystallizer cooler and most of the $CuCl_2$ crystallizes. This cooler could be similar to the scrapped surface crystallizer manufactured by Armstrong Engineering. After exiting the cooler, the solid $CuCl_2$ is collected in the bottom of a hydroclone. The solid $CuCl_2$ is then fed to the hydrolysis reactor. The overflow of the hydroclone is recycled back to the anode feed tank through a pressure let down valve.

Cathode and Hydrogen Generation Section: Aqueous HCl is pumped (24 bar) from the cathode feed tank to the cathode. At the cathode the H^+ ion is reduced to H_2. The chloride ion migrates across the electrolyzer membrane as described above. Water and unreacted HCl are then heated to 106°C and flashed to atmospheric pressure. The vapor from the flash (which contains less than 1mole % HCl) is then sent to the direct heat exchanger and superheated to 400°C. It is then sent to the hydrolysis reactor. The liquid stream from the flash is recycled back to the cathode feed tank.

Hydrolysis/Oxychloride Decomposition

The solid product from the hydrolysis reactor (*2*) goes to the oxychloride decomposition reactor (*3*). Thermal energy is used as the final preheat agent of all streams to the hydrolysis reactor and to decompose the Cu_2OCl_2 in the oxychloride decomposition reactor. The reactors are sized using residence times observed in the laboratory. The effluent from the oxychloride decomposition reactor is a molten salt. The heat from the molten salt is recovered in a direct heat exchanger (see below).

Hydrolysis reactor: The hot, pressurized (24 bar) $CuCl_2$ is sprayed into a superheated (400°C) steam environment at 0.25 bar where it forms a free jet. As the jet expands it aspirates the superheated steam into the jet resulting in high mass and heat transfer between the $CuCl_2$ in the jet and the steam. The $CuCl_2$ is converted to Cu_2OCl_2 and HCl.

The HCl and unreacted steam exit the hydrolysis reactor to be cooled in feed pre-heaters and fed to the cathode of the electrolyzer. A steam ejector and the volume contraction of the hydrolysis steam/HCl effluent pulls the vacuum on the hydrolysis reactor as it condenses. Dry, free flowing solid Cu_2OCl_2 accumulates at the bottom of the hydrolysis reactors. The solid copper oxychloride flows by gravity through an L valve to the oxychloride decomposition reactor. The hydrolysis reactor is elevated relative to the oxychloride decomposition reactor. The static head of the solids in the L valve provide the driving force to move the solids from the hydrolysis reactor to the oxychloride reactor.

The hydrolysis reactor is very similar to a spray roaster used in the steel industry to recover HCl from $FeCl_2$ generated by the steel pickling process. The flow of solid Cu_2OCl_2 from the hydrolysis reactor to the oxychloride decomposition reactor is similar to the flow of cracking catalyst in a fluid bed catalytic reactor in a refinery.

Oxychloride decomposition reactor. In the oxychloride decomposition reactor, the Cu_2OCl_2 is heated to 550°C. Between 450 and 550°C, the Cu_2OCl_2 decomposes to oxygen and molten CuCl. The oxygen leaves the oxychloride decomposition reactor as a gas and the molten CuCl spills over the weir. The stand pipe below the oxychloride decomposition reactor can be fluidized with air if required to assure smooth flow into the oxy-decomposition reactor. Air is aspirated into the oxygen stream using a venturi valve to cool and dilute the oxygen.

The Direct Heat Exchanger. The heat in the molten CuCl stream is recovered in a heat exchanger followed by a direct heat exchanger. The molten CuCl enters the direct heat exchanger at 440°C (after it cooled down in a shell and tube heat exchanger from 550 to 440°C) and is atomized by a spinning disc. As the molten CuCl cools it heats the vapor stream from the cathode flash. The vapor stream is then fed to the hydrolysis reactor. During the cooling the CuCl is granulated. The granulated CuCl is fed back to the anode feed tank via a screw feeder or via gravity flow. The direct heat exchanger is very similar to the Bateman Granulation system which is used to make granulated steel or slag.

Electrolyzer Design

The proposed electrolyzer combines individual electrolyzer cells consisting of a membrane and anodic/cathodic compartments using a modular design of stacks and modules to attain the necessary hydrogen production throughput. The ion-exchange membrane is located in the middle of the cell and is sandwiched between the two electrodes, the anode and the cathode. Both the anode and cathode are porous, carbon/carbon-felt electrodes typically used in PEM fuel cell design. In the CuCl electrolyzer, the cathode has an electrocatalyst layer to promote the hydrogen generation reaction whereas the anode does not need this layer. Moving outwards from the electrode is the bipolar plate, also called a flowplate, which acts as a channel for gas and electrolyte flow into and out of the anodic and cathodic compartments. In a multiple-cell configuration, the bipolar plate would have flow channels on each side.

In the proposed modular electrolyzer design, multiple cells are combined to make an electrolyzer stack. The stacks are then arranged into core modules of 2 or more stacks each in order to optimize the reactant/product flow, power, and operational and maintenance requirements. For the CuCl hybrid thermal process, three electrolyte flow trains have been proposed. Thus, the electrolyzer stacks should be grouped in such a way to make three core modules with a single input and output stream each for the anode and cathode electrolyte flows and with the proper amount of stacks per module to generate the hydrogen throughput. For a single cell area of 3 m^2, 300 cells/stack and 10 stacks/module, 6 modules (2 per process train) are required to achieve the 27,000 m^2 to produce 125,000kg of hydrogen/day. The 6 modules are contained in 9 pressure vessels.

3.1.1. H2A Economic Results

DOE has developed the H2A spread sheet [1] to evaluate potential hydrogen production projects. The H2A is a discounted cash flow analysis. It calculates the cost of producing hydrogen if a 10% return on equity is required.

The capital investment in the electrolyzer was estimated separately at $38.4 MM. Using Capcost software, the total capital investment (including offsites) for the CuCl plant is $105 MM. The number of operators required to operate the plant is 80. This number is high because of the 3 (crystallizer, hydrolysis and oxy-decomposition reactor) process steps which involve solids. The cost of energy from a nuclear source is estimated at $20/MW-hr (thermal) and $60/MW hr (electrical). All estimates are in 2005 dollars.

When this investment, labor force and cost for energy (both thermal and electrical) are plugged into the H2A spread sheet, the estimated cost of hydrogen is $3.07/kg. A number of sensitivities were run on the economics. Depending on the sensitivity tested, the cost of hydrogen can range from $3.60 to 2.80/kg. As would be expected the cost is most sensitive to the plant capital cost, the operating factor and the cost of energy (both electrical and thermal). The cost is relatively insensitive to the size of the labor force.

3.2. Model of the Hydrolysis Reaction

The hydrolysis reactor was modeled with Aspen Plus with a feed of 100 kmol/hr of $CuCl_2$ and various amounts of water. Figure 3a shows the effect of temperature on the Cu_2OCl_2 yield as a function of the steam-to-copper (S/Cu) ratio as calculated by Aspen Plus™ model. An excess of steam is required for achieving high yields of Cu_2OCl_2. For instance, to achieve the maximum yield of 50 mol of Cu_2OCl_2 below 400°C, a steam-to-copper ratio greater than 17 is needed. Below 400°C, the model also predicts that the thermal decomposition reaction of $CuCl_2$ to CuCl occurs to a small extent, as seen in Figure 3b.

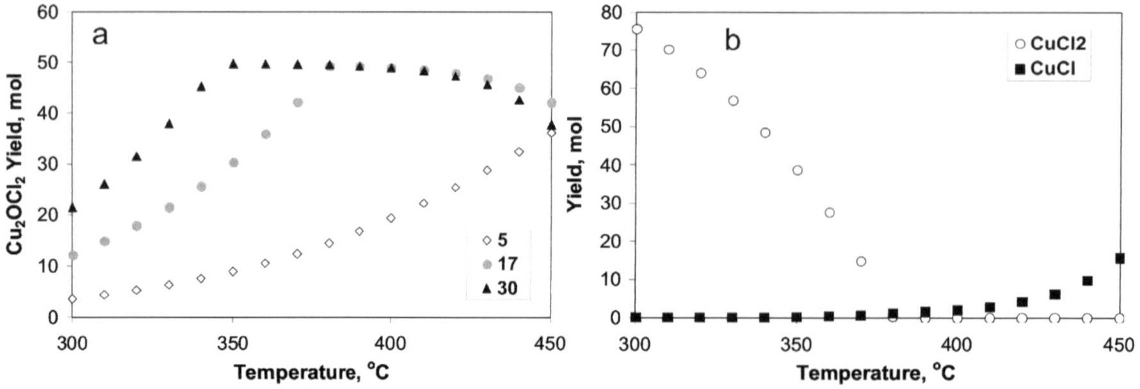

Figure 3. Effect of temperature on the Cu_2OCl_2 yield as a function of S/Cu ratio (a) and CuCl and $CuCl_2$ yields as a function of temperature (S/Cu=17) (b), calculated by Aspen Plus model (Ar=0).

3.3. Results from the Hydrolysis Experiments

The conditions varied were the following: flow rate of argon through the pneumatic nebulizer, the types of flow designs (co-current versus counter-current) and the types of atomizers. Scanning electron microscope (SEM) was used to measure the particles size of the solid product obtained. X-ray diffraction technique (XRD) was performed for identification of CuCl, Cu_2OCl_2 and $CuCl_2$. Analyses of CuCl were performed using a wet chemistry method by the Analytical Chemistry Laboratory at Argonne.

In Figure 4a, the effect of varying Ar flowrates from 200 to 400 mL/h through the nebulizer at *co-current* design is shown. Most of the reaction product is unreacted $CuCl_2$. We attribute this result to poor mixing and poor mass transfer. The amount of CuCl is small but increases with the Ar flowrate, as expected due to Le Chatelier's Principle. All flowrates show very little conversion into Cu_2OCl_2. All the peak intensities attributed to $CuCl_2$ and $CuCl_2{\cdot}2H_2O$ were lower at higher Ar flowrates. This suggests that at higher Ar flowrate the droplets are finer and hence the size of the solid particles is smaller.

In Figure 4b, the results are shown for the same type of experiments as in 2a but the flow is *counter-current*. These results suggest that counter-current flow gives more turbulence and better mixing. Very fine particles of black color, which is the color of Cu_2OCl_2, were found at the bottom of the reactor. Further, more Cu_2OCl_2 is formed relative to the amount of unreacted $CuCl_2$ in 4b. As the Ar flow increases, the solution droplets become finer and there is greater dispersion. We attribute the larger conversion to improved mass transfer.

Figures 5a and 5b show XRD patterns obtained from the nebulizer and the ultrasonic nozzle for the same conditions. The XRD pattern of the solid product obtained using the ultrasonic nozzle shows mostly Cu_2OCl_2, little unconverted $CuCl_2$ and much less CuCl compared to the pneumatic nebulizer. Analyses of the product showed less than 4 wt% of the undesired CuCl obtained using the ultrasonic nozzle. The difference in the results obtained with the two types of nozzle may come from: *(i)* the droplet size and *(ii)* the velocity at the tip of the atomizer. SEM pictures reveal the solid particles after reaction using the ultrasonic nozzle to be in the range of 2-3 μm with a homogeneous size distribution (Figure 6), which also corresponds to the mean size of liquid droplets (25 μm) according to the manufacturer.

The velocity of the gas/liquid at the tip of the ultrasonic nozzle is smaller than that of the pneumatic nebulizer because it does not use an inert gas and the orifice diameter is larger (510 μm versus 190 μm for the pneumatic nebulizer). Thus the terminal velocity is reached at a shorter distance in the reaction zone from the atomizer and the residence time is longer with the ultrasonic nozzle, resulting in a better heat and mass transfer.

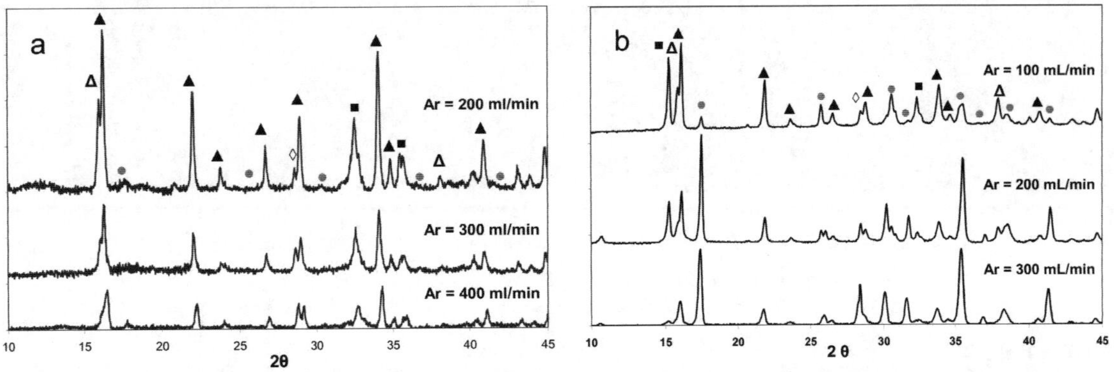

Figure 4. XRD pattern of solid products. Effect of different Ar flow rates using co-current (a) and counter-current (b) flow designs. Cu_2OCl_2 (●), $CuCl_2{\cdot}2H_2O$ (▲), $CuCl_2$ (Δ), CuCl (◊), Cu(OH)Cl (■).

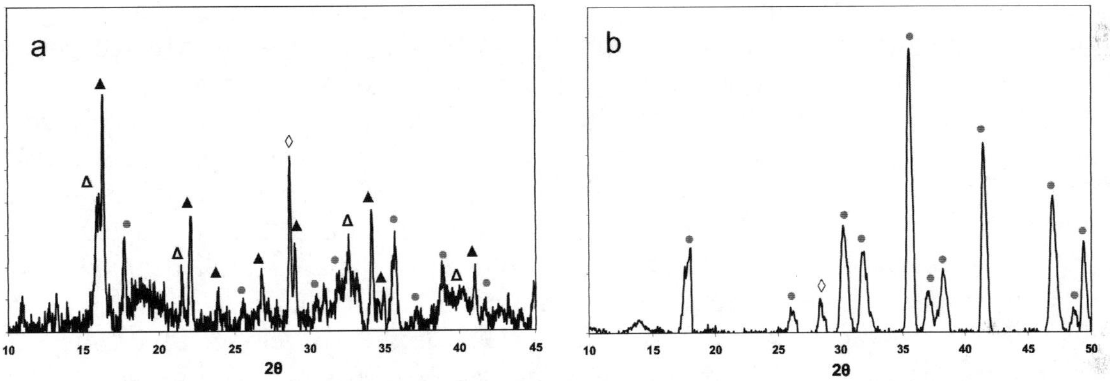

Figure 5. XRD pattern of solid products. Effect of different atomizers: pneumatic nebulizer (a) and ultrasonic nozzle (b). Cu_2OCl_2 (●), $CuCl_2{\cdot}2H_2O$ (▲), $CuCl_2$ (Δ), CuCl (◊), Cu(OH)Cl (■).

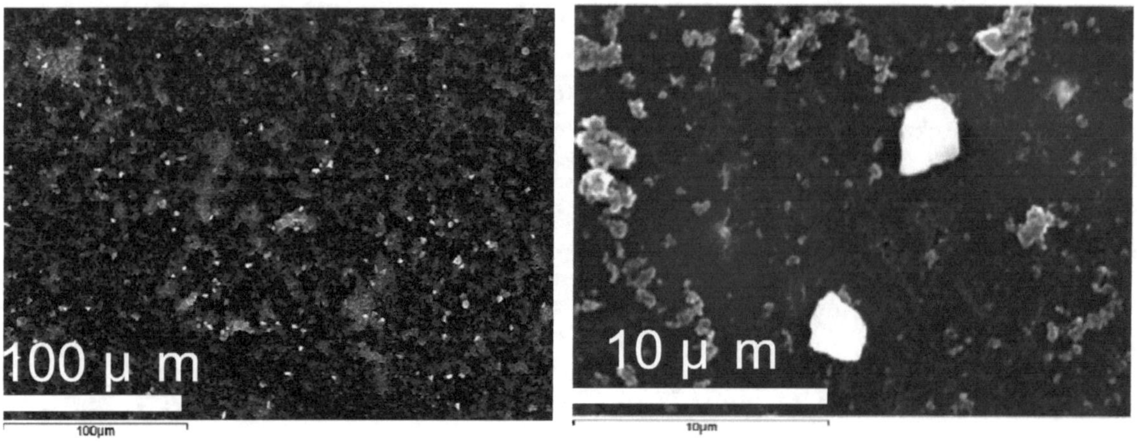

Figure 6. SEM pictures of particles of solid products (white) embedded in epoxy after hydrolysis reaction using the ultrasonic nozzle

4. Conclusions

A conceptual process design based on the 3 reaction Cu-Cl cycle has been developed to produce 125 MT of hydrogen /day. The process is based on an Aspen Plus simulation which provides mass and energy balances for the process. An integrated heat exchange network for the conceptual process captures process heat. External heat and utility requirements are 191MWt and 100.5 MWe respectively. Defining efficiency as energy out divided by energy in results in estimated efficiency of around 39%. The voltage assumed for the electrolyzer is 0.7V which is for ambient conditions. We plan to plan to run the electrolyzer at 80°C, so the voltage could be lowered [5], thus the efficiency should go up. Capital requirements for the conceptual process were estimated using Capcost software. The chemical plant investment including the electrolyzer is $105MM (2005$). Operating costs were also estimated and an H2A analysis was preformed. The estimated cost of producing hydrogen is $3.07 with a range from $3.60 to 2.80/kg depending capital investment, amount of equity financing, and the cost of thermal and electrical energy. The results of the economic analysis are guiding the further development efforts, both experimental and modeling. Efforts are being focused in areas which will improve process efficiency and reduce capital investment.

During the hydrolysis of $CuCl_2$ with the pneumatic nebulizer, counter-current flow designs gave higher yields of Cu_2OCl_2 than co-current design. However, the latest results indicate that the best results are obtained when using an ultrasonic nozzle in the co-current design. Very fine and homogenous particles result, thereby providing good mass and heat transfer.

5. Acknowledgments

Argonne National Laboratory's work was supported by the U.S. Department of Energy, Office of Nuclear Energy, Nuclear Hydrogen Initiative Program, under contract DE-AC02-06CH11357. The authors gratefully acknowledge the assistance provided by Dr. Paul Mathias, who developed the physical properties database for the Aspen Plus simulation and completed the sensitivity analyses.

6. References

1. www.hydrogen.energy.gov/h2a_analysis.html.
2. Turton, R., Baillie, R.C., Whiting, W.B., Shaeiwitz, J.A., Analysis, Synthesis and Design of Chemical Processes (2nd Ed), Upper Saddle River, NJ: Prentice Hall, 2003.
3. Ferrandon, M., Lewis, M.A., Tatterson, D.F., Nankani, R.V., Kumar, M., Wedgewood, L.E., Nitsche, L.C., *The hybrid Cu-Cl thermochemical cycle. I. Conceptual process design and H2A cost analysis. II. Limiting the formation of CuCl during hydrolysis*, Extended Abstract submitted for the NHA Annual Hydrogen Conference 2008, Sacramento, CA, March 31-April 4 2008.
4. ASPEN PLUS User Guide; Aspen Technology, Inc.: Cambridge, MA, 1988.
5. Summer, W.A., Corgnale, C., Status of Hybrid Sulfur Process for STCH Down-select, DOE STCH Technical Progress Meeting, Argonne National Laboratory, Chicago, IL, August 27-28, 2008.

DESIGN OF RECUPERATIVE SULFURIC ACID DECOMPOSITION REACTOR FOR HYDROGEN GENERATION PROCESSES

Sarah M. Connolly[1], Edward J. Lahoda[1], and David F. McLaughlin[1]
[1]Westinghouse Electric Company, LLC. Pittsburgh, PA

Executive Summary

The Hybrid Sulfur Process (HyS), as well as similar sulfur cycles for the production of nuclear hydrogen, requires the decomposition of sulfuric acid into sulfur dioxide, oxygen, and water at temperatures above 800°C and a pressure near 9 MPa. The design of a reactor for this process presents numerous challenges in terms of maintaining small pressure differentials and utilizing currently available materials of construction. This paper focuses on design calculations for a recuperative decomposition reactor for use in the production of hydrogen. The decomposition reaction takes place within individual tubes of a multi-tube reactor that are is fed liquid H_2SO_4 at 50 to150°C.

Previous calculations involved modeling to produce a detailed temperature-enthalpy (T-H) diagram for the H_2SO_4-SO_3-SO_2-O_2-H_2O system of the evaporation and decomposition processes for pinch analysis. Previous detailed modeling and analysis has been performed for the Westinghouse design by Connolly, et al.[(2)] Current analysis involves detailed modeling and sizing of the recuperative concept. This design phase entailed detailed heat transfer analysis and sizing of the reactor. Calculations were performed with a multi-tube reactor design in which the outer shell of the reactor contained structured packing to facilitate heat transfer on the shell side. This resulted in a design consisting of a total of 750 tubes at 6.7 meters in length contained in two vessels with an inner diameter of 3.7 meters.

1. Introduction

1.1 Background

The sulfur cycles, including the Hybrid Sulfur Process and the Sulfur-Iodine Process, are a group of processes designed to generate hydrogen via electrolysis and thermochemistry. The Hybrid Sulfur Process (HyS), or the Westinghouse Sulfur Process (WSP), uses both thermal and electrical energy from a Pebble Bed Modular Reactor (PBMR) to produce hydrogen. The thermal energy is provided by helium that is used as a coolant in the nuclear reactor and is transferred via an intermediate heat exchange loop to a second helium stream. The thermal energy from the intermediate loop is used to decompose H_2SO_4 into SO_2, O_2, and H_2O. Oxygen is then separated from the stream and the water-SO_x mixture is electrolyzed to generate H_2 gas and H_2SO_4 via the reaction:

$$2\,H_2O + SO_2 \rightarrow H_2 + H_2SO_4 \qquad \textbf{Equation 1}$$

The H_2SO_4 can then be recycled back into decomposition reactor. A diagram of the HyS process is shown in Figure 1.

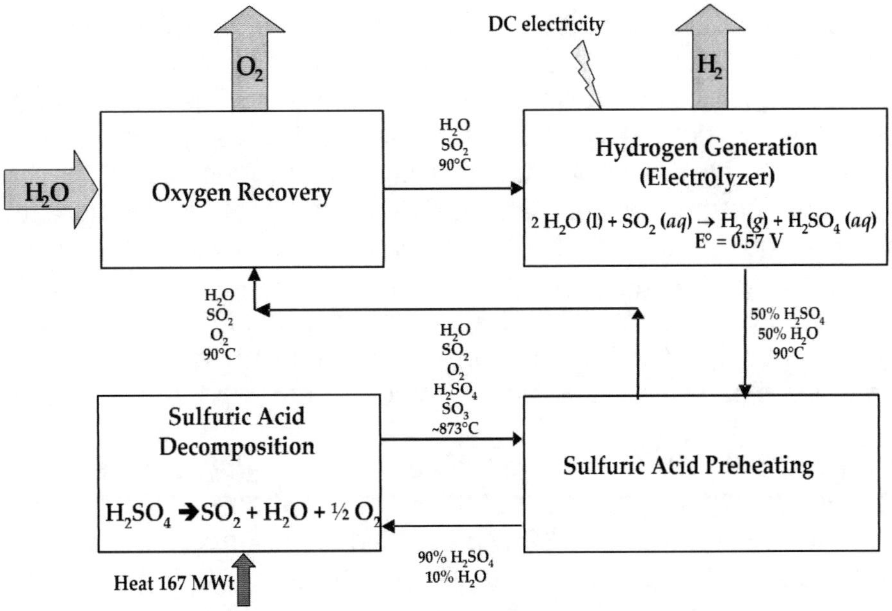

Figure 1 – Diagram of HyS Process

1.2 Design Challenges

All of the sulfur cycle processes require the decomposition of concentrated sulfuric acid into sulfur dioxide, oxygen, and water at temperatures exceeding 800°C. Additionally, the lower condensation temperature of the decomposition products relative to the feed stream requires operating pressures near 9 MPa. The high temperatures and pressures in combination with the extremely corrosive nature of the materials present considerable challenges in designing a reactor system.

It is possible to overcome the challenges presented by both the high temperatures and the corrosive nature of the H_2SO_4 by using a ceramic material such as silicon carbide (SiC) or alumina (Al_2O_3). However, the use of ceramics then presents the problem of creating seals between the low-temperature metal components and the high-temperature ceramic components of the reactor. Plastic materials, such as Teflon, are only allowable at temperatures up to 300°C. The use of metal gaskets would create issues of corrosion and possible loosening under cyclic operating conditions. Additional issues include maintaining small pressure differentials between the helium side of the reactor and the sulfuric acid side as well as creating a design that is amenable to maintenance.

2. Westinghouse Composite Reactor

2.1. Westinghouse Design Concept

The Westinghouse solution to this design challenge is a top-fed, multi-step reactor that pre-heats, concentrates, and decomposes sulfuric acid. The reactor will operate under moderate pressure differentials (<1 MPa) between the heated gas and liquid sides. Ceramic material will be utilized for the high temperature and heat transfer portions of the reactor and lined metal in the low temperature distribution headers of the reactor. All seals are located in low temperature portions of the reactor, allowing for the use of Teflon or a similar material. A general schematic of the composite reactor is shown in Figure 2. The design allows for a relatively low cost of construction by using carbon or stainless steel for the pressure boundary of the vessel.

The decomposition tubes consist of a ceramic bayonet tube with an inside ceramic support tube, as shown in Figure 3. The bayonet tubes are contained in a vessel with internal insulation and structured packing to facilitate heat transfer from the helium to the bayonet tube. Inside the bayonet tube, a double-walled insulated ceramic support tube surrounded by a ceramic tube to create a gas downflow annulus sits on the bottom of the bayonet tube and supports a ceramic liquid catch ring with sloped sides and a hole in the middle. A ceramic gas distribution plate then sits on top of the liquid catch ring, which in turn supports ceramic packing material. On top of the packing material is a set of ceramic gas riser tubes filled with packing. At the bottom of the bayonet tube, ceramic packing material sits in the annulus between the bayonet and gas down flow ring and supports a catalyst to be used in the decomposition reaction. The entire assembly is made of ceramic, such as SiC, and is handled as a single unit.

The incoming liquid H_2SO_4 feed flows between the ceramic gas riser tubes where it is preheated by the upward flowing gases from the decomposition products. The liquid is directed downward into the liquid catch ring via the gas distribution plate where it flows down through the support tube. At the bottom of the support tube, the liquid can enter into the bayonet via holes or perforations in the support tube. The H_2SO_4 is further concentrated to 85-90 wt% then vaporizes and is demisted by the ceramic packing as it flows upward. As the H_2SO_4 vaporizes, it is decomposed into SO_3 and H_2O. Upon reaching the catalyst, the SO_3 is further decomposed into SO_2 and O_2. At the top of the bayonet tube, the gas will flow downward into the inner annulus where heat is recuperated into the incoming feed stream.

The primary advantages to the Westinghouse design lay in its amenability to assembly, inspection, and maintenance as well as the requirement for only one seal location. Additionally, the seal is not under a hydrostatic head that could result in a leak into the helium side.

2.2 Sandia National Laboratory Concept

The Sandia National Laboratory (SNL) solution to the design challenge is a reactor in which the decomposition reaction takes place within individual tubes of a multi-tube reactor that are pump-fed 78 wt% H_2SO_4 at 50 to 150°C. As in the Westinghouse design, ceramic material will be utilized for the high temperature and heat transfer portions of the reactor and lined metal in the low temperature portions of the reactor. Three seal locations are required, and all are located in low temperature portions of the reactor, allowing for the use of Teflon or a similar material. A general schematic of the decomposition reactor is shown in Figure 4. The design also allows for a relatively low cost of construction by using carbon or stainless steel for the pressure boundary of the vessel.

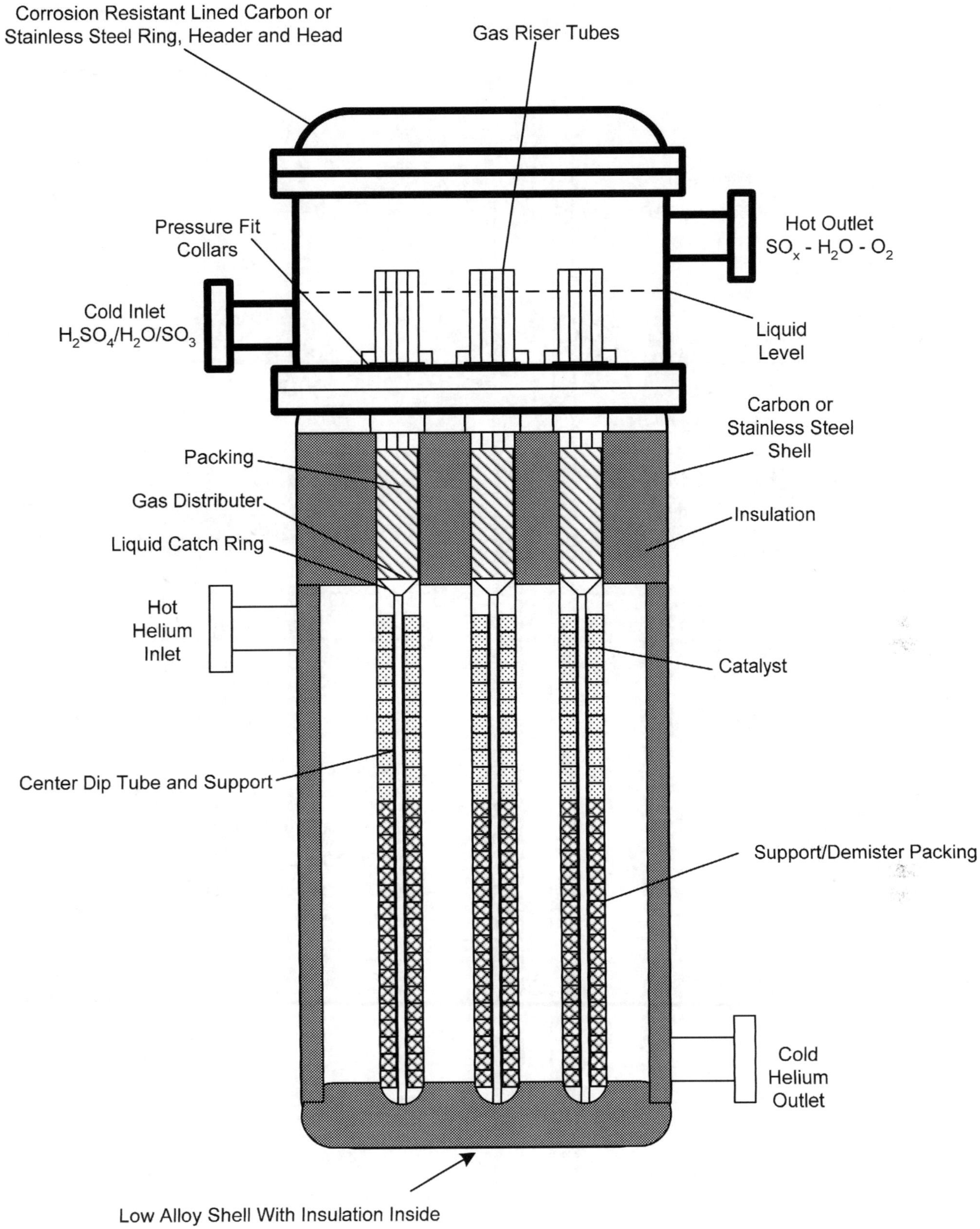

Figure 2 – Overall Schematic of Westinghouse Composite Reactor

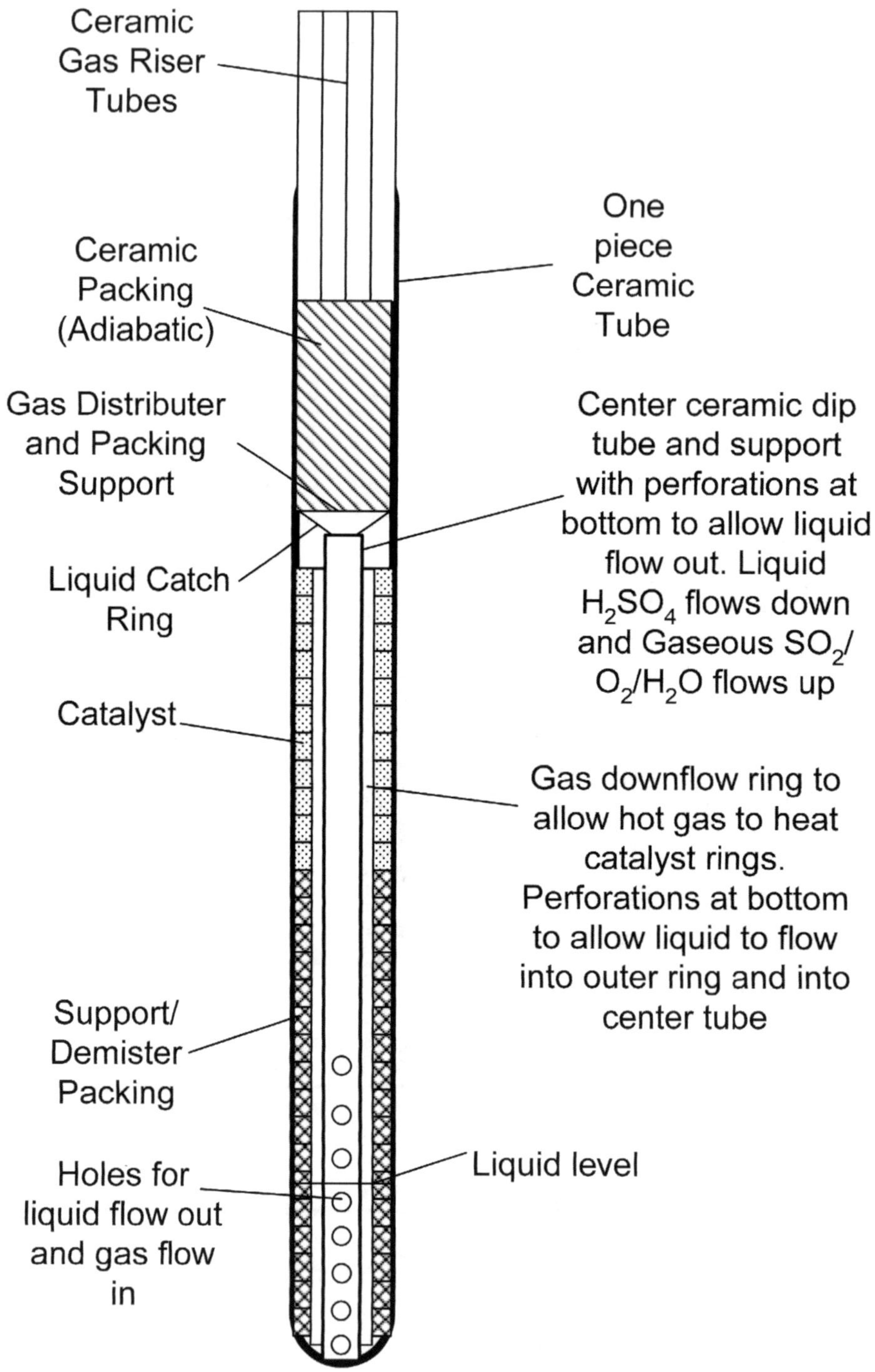

Figure 3 - Detail of Individual Westinghouse Composite Operation Tube

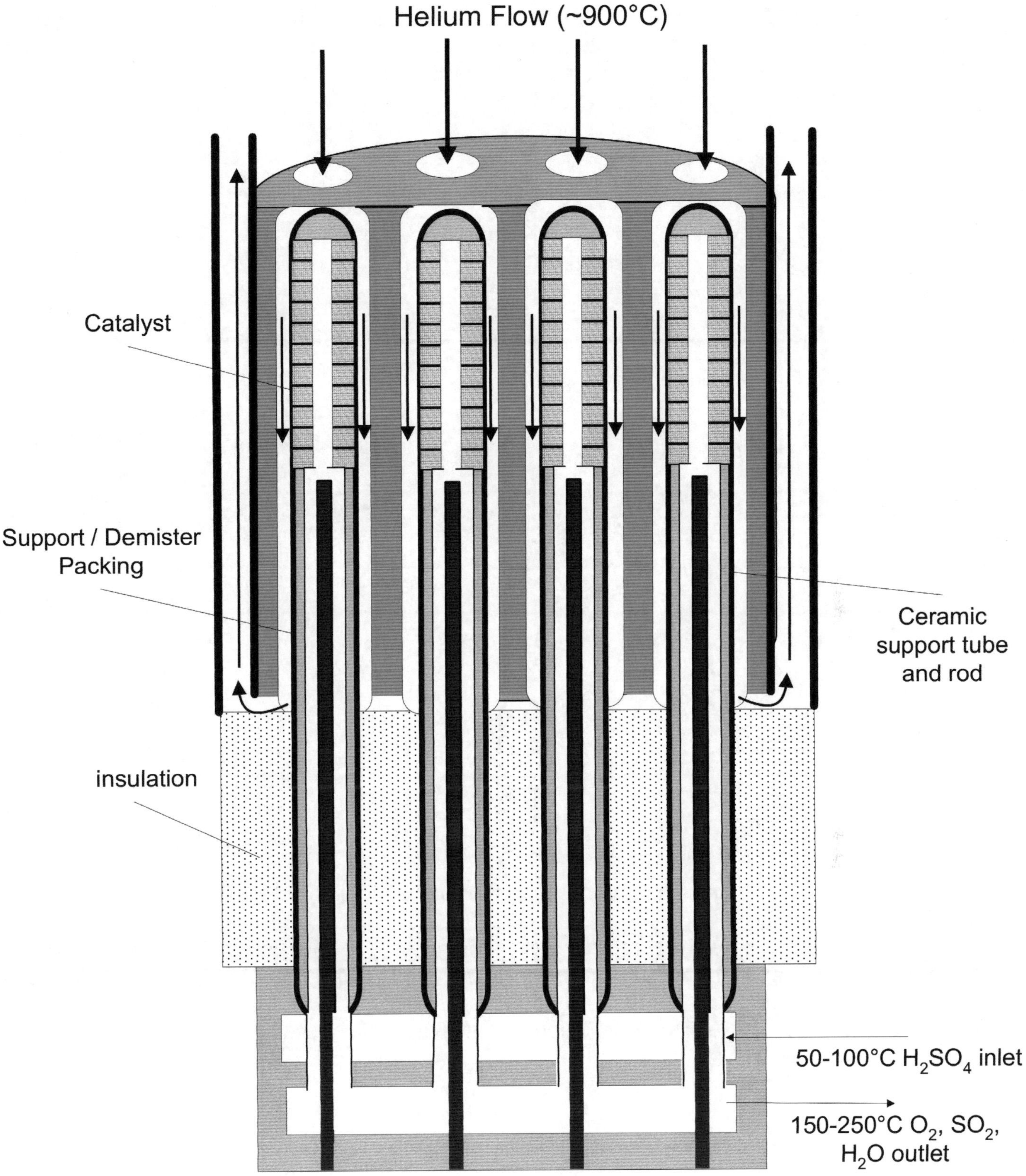

Figure 4 – Overall Schematic of H_2SO_4 Decomposition Reactor

The decomposition tubes consist of a ceramic bayonet tube with an inside ceramic support tube. A rod in the center of the exit tube creates an inner annulus to facilitate heat transfer, as shown in Figure 5. The bayonet tubes are contained in a vessel with internal insulation and structured packing to facilitate heat transfer from the helium to the bayonet tube. The entire assembly is made of ceramic, such as SiC.

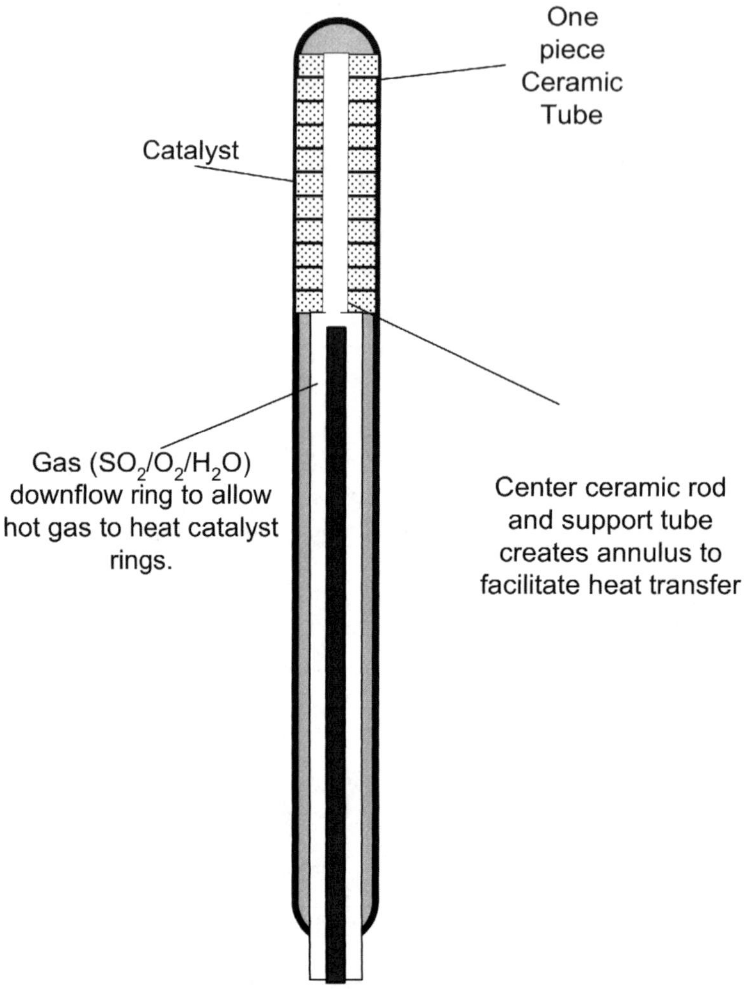

Figure 5 – Detail of Individual SNL Design Tube

The primary drawback to the SNL design is the necessity for three seals – a helium/reactant seal and two seals to maintain separation of the inlet and outlet streams (see Figure 6). The temperature in these areas is less than 300°C, therefore various commercially available seals can be utilized. Additionally, active cooling coils could be used for further cooling if desirable.

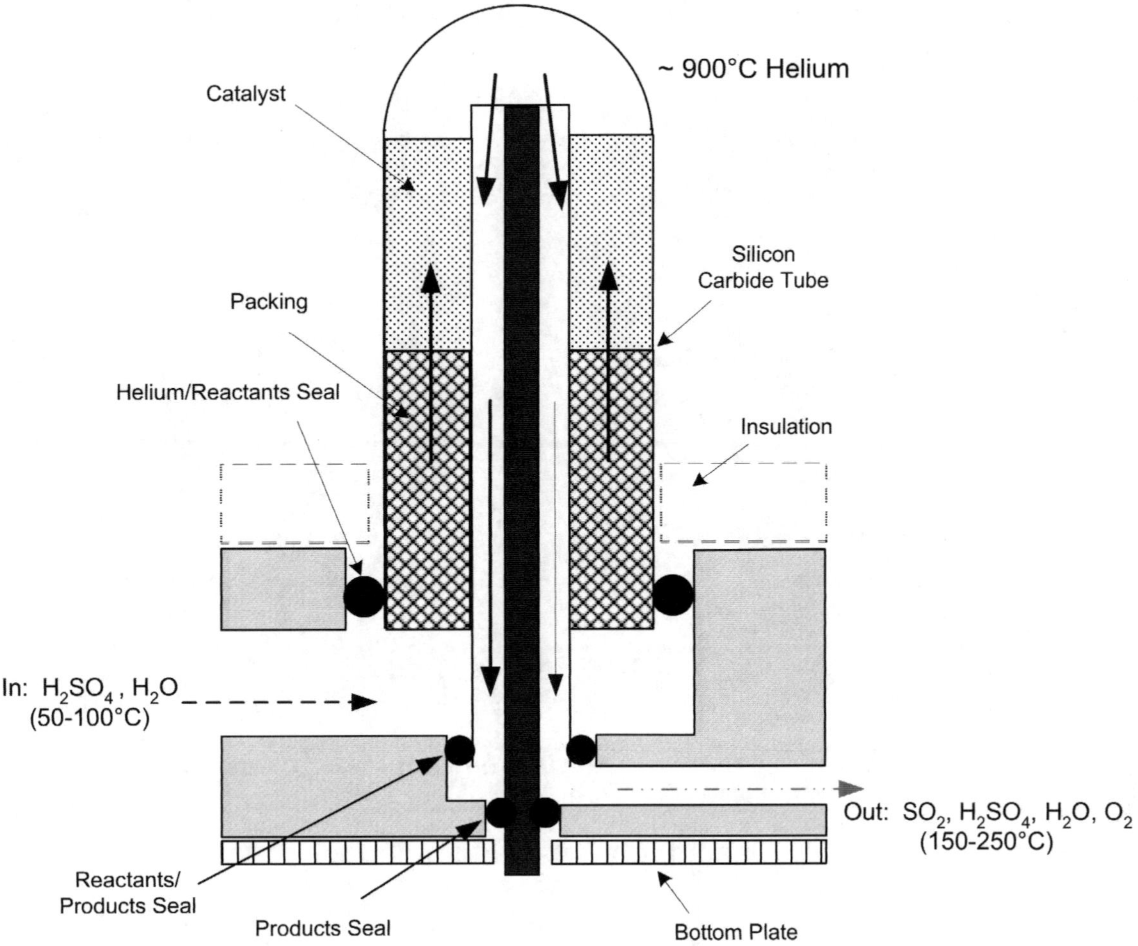

Figure 6 – Detail of Seal Locations

3. Development of Thermodynamic Model

3.1. T-H Diagram

Development of an accurate thermodynamic model was necessary for detailed modeling of the decomposition reactor. Such development is difficult due to the scarcity of vapor pressure data for highly concentrated H_2SO_4 systems. The highest temperature vapor pressure data available are those of Wüster[(1)], which reach a maximum temperature of 480°C for a range of concentrated solutions. Detailed development of the thermodynamic model is discussed in Connolly, et al.[(2)] The final T-H diagram shown in Figure 5 was developed using ChemCAD data up to the ASPEN predicted bubble point of 492°C, ASPEN data from 492°C to the dew point at 584°C, extrapolated CEA data from 492°C to 607.6°C and CEA data above 607.6°C. The merging of the data at the bubble and dew points is markedly smooth, indicating good correlation between the three data sources. Calculation of the T-H line for the helium is straightforward. Figure 7 confirms that there is no pinching between the helium and the sulfuric acid systems.

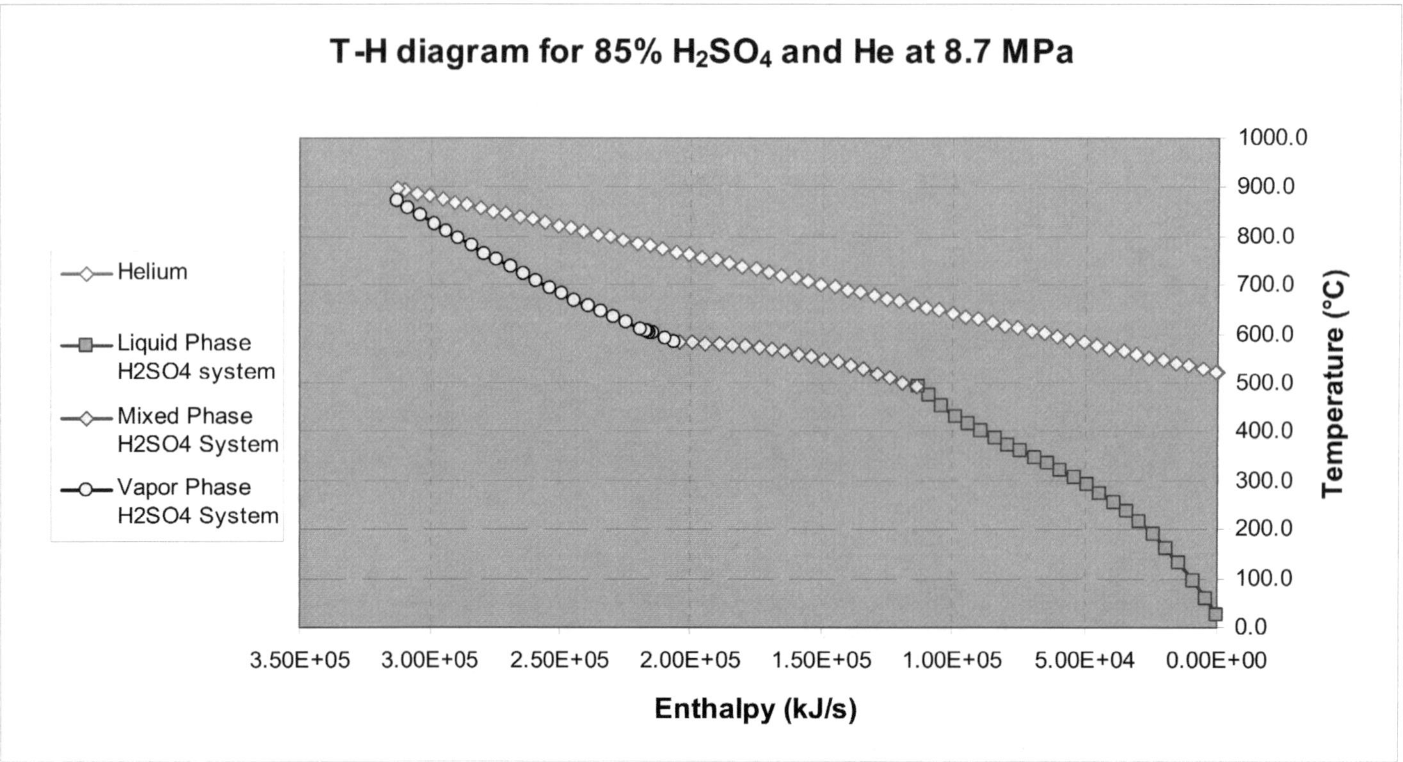

Figure 7 - T-H Diagram for Decomposition Reactor at 8.7 MPa

4. Decomposition Reactor Model

4.1. Design Equations

Previous detailed modeling and analysis has been performed for the non-recuperative Westinghouse design by Connolly, et al.[2] Current analysis involves detailed modeling and sizing of the recuperative concept by Westinghouse in order to form a basis for comparison. The model for the decomposition reactor is set up to balance the enthalpy and the heat transfer between the helium side, the outer annulus in which the decomposition reactions take place, and reaction products flowing downward through the inner annulus. The system consists of six total possible variable temperatures – the inlet and outlet temperatures for the helium side, the outer annulus, and the inner annulus. The temperature at the top of the inner (T_{inIA}) and outer (T_{outOA}) annuli must be the same and is set at 872°C by the WSP flow sheet. The inlet temperature of the helium stream (T_{inHe}) is also set by the WSP flowsheet. Thus, the three unknown temperatures – outlet temperature of inner annulus (T_{outIA}), outlet temperature of the helium (T_{outHe}), and the temperature of the feed (T_{inOA}) – serve as the variables for the system. Hence, three equations relating these temperatures via enthalpy and heat balances are required to model the system. The flow of helium on the shell side and the reaction products through the inner annulus are both downward and the feed in the outer annulus flows upward, which results in countercurrent flow between the helium and the outer annulus and also between the outer annulus and inner annulus, as shown in Figure 8.

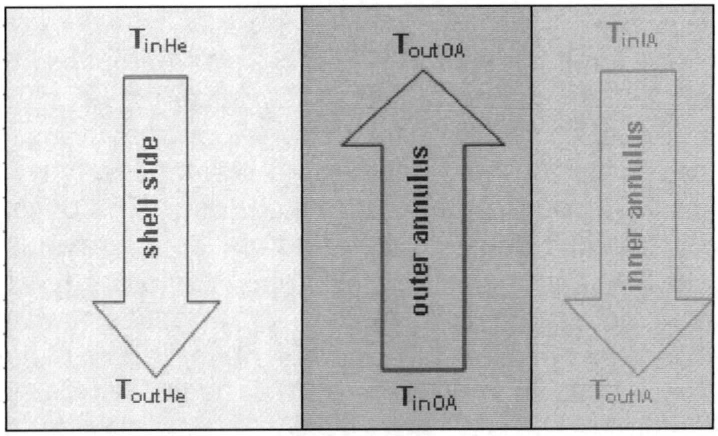

Figure 8 - Flow Pattern Through Decomposition Reactor

The first design equation (Equation 2) consists of an enthalpy balance on all three systems. The enthalpy leaving the helium must be exactly equal to the enthalpy that is gained by the H_2SO_4 system in the outer and inner annuli assuming that the overall operation is adiabatic. Enthalpy increases with temperature and the temperature of the system decreases with the direction of flow for the helium and the inner annulus but increases with the direction of flow in the outer annulus. The change in enthalpy, ΔH, is defined as $H_{out} - H_{in}$, thus the enthalpy change on the shell side and in the inner annulus will be negative and the enthalpy change in the outer annulus will be positive. The enthalpy balance is thus:

$$\dot{m}_{He}\Delta H_{He} + \dot{m}_{H2SO4}(\Delta H_{OA} + \Delta H_{IA}) = 0$$ **Equation 2**

The remaining two equations are provided by balancing the heat transfer into and out of the inner and outer annuli. The enthalpy change in the outer annulus is balanced by the heat transfer into the outer annulus from the helium and the inner annulus, as shown in Equation 3 where $U_{OA/IA}$ and $U_{He/OA}$ are the overall heat transfer coefficients between the inner and outer annuli, and the helium side and the outer annulus, respectively. Heat transfer correlations were used as described in Connolly et al.[(2)] A_1 corresponds to the area available for heat transfer between the shell side and the outer annulus, and A_2 corresponds to the area available for heat transfer between the outer and inner annuli. The standard log mean temperature differences for each system are represented by ΔT_{LM}.

$$\dot{m}_{H2SO4}\Delta H_{OA} = U_{OA/IA}A_2\Delta T_{LMOA/IA} + U_{He/OA}A_1\Delta T_{LMHe/OA}$$ **Equation 3**

In the inner annulus, the change in enthalpy is simply balanced with the heat transfer from the inner annulus into the outer annulus, as shown in Equation 4.

$$\dot{m}_{H2SO4}\Delta H_{IA} = -U_{OA/IA}A_2\Delta T_{LMOA/IA}$$ **Equation 4**

The values for the change in enthalpy are computed simply by regressing the data for the T-H diagrams to give temperature dependent functions of H. Similarly, the overall heat transfer coefficients are simply functions of the physical properties of the materials present in the system (which are temperature dependent) and the tube geometries.

4.2. Model Development

The current model for the decomposition reactor partitions the total required reactor tube length into 20 equal sections and balances the three previously stated design equations for each section. The model allows the user to vary tube diameters, clearances, and lengths as well as mass flow rates until the heat transfer requirements are met and pressure drops are kept below approximately 20 pounds per square inch (psi). The optimization limits for various tube geometries were obtained from either literature values or previous calculations.

The primary optimization parameter was the minimization of the total number of required tubes while still maintaining acceptable pressure drop losses and necessary heat transfer. The number of tubes was chosen as the primary optimization variable because initial cost estimates show that the internal components of the reactor will comprise the majority of the total reactor cost due to the necessity for corrosion resistant materials at high temperatures and pressures.

4.3. Optimization Calculations

To increase the heat transfer on the helium side of the reactor, it was proposed that using structured ceramic packing rather than the standard baffled approach would increase the heat transfer coefficient and eliminate the need for using a correction factor in the log-mean temperature difference. The assumed structured packing used for calculation purposes was Flexeramic® Type 38. This type of packing is typically used in mass transfer applications, however the effects on heat transfer and pressure drop are relevant.

The final optimized model utilizes 750 bayonet tubes distributed within two vessels with inner diameters of 3.2 meters, as shown in Table 1. The height of the tubes is 7.4 meters. The pressure drops on the shell side, outer annulus, and inner annulus were 18.9 psi, 9.2 psi, and 1.4 psi, respectively.

The temperature profile for the decomposition reactor is shown in Figure 9. Note that the zero point on the x-axis corresponds to the top of the tube. The temperatures in the inner and outer annuli are the same at the top of the reactor. Note that both the inlet feed temperature and the reaction products temperature are below 300°C at the bottom of the reactor. External heat transfer via the helium is only utilized until the helium reaches 500°C, as required by the other components of the HyS process.

4.4. Pressure Drop Calculations

Various methods of predicting the pressure drop through the different sections of the decomposition reactor were required. Specific equations for the pressure drops are available in Appendix A. The pressure drop through the bayonet portion was calculated using the Ergun equation with characteristic length and velocity relations as reported in Mills[(3)]. The pressure drop in the support tube was calculated using a friction factor correlation for the transitional regime and a pressure drop equation for flow through circular pipes as reported in Welty[(4)].

The method of calculating pressure drop on the shell side depends on the design approach. For the packed design, the method for dry pressure drop in structured ceramic packing detailed by Uresti-Melendez and Rocha[(5)] was utilized. The total pressure drop for the shell side, outer annulus, and inner annulus as a function of tube length is shown in Figure 10.

Table 1 - Optimization Table for Recuperative Decomposition Reactor

Inside shell diameter	3.7	m
number of bayonet tubes required	750	
number of vessels	2	
total length of tubes	6.7	m
clearance between tubes	0.035	m
ID of support tube (inner wall)	0.020	m
thickness of support tube	0.004	m
insulation thickness	0.010	m
OD of support tube (outer wall)	0.056	m
distance in inner annulus	0.0025	m
ID of inner annulus	0.061	m
thickness of inner annulus	0.006	m
OD of inner annulus	0.073	m
distance in outer annulus	0.030	m
ID of outer annulus	0.103	m
thickness of outer annulus	0.021	m
OD of outer annulus	0.144	m
Pitch/OD outer annulus	1.242	
Catalyst void volume (ε)	50%	
catalyst/packing particle diameter (d_p)	0.010	m
heat transfer area on outside inner annulus	0.08	m^2
area available on shell side (per finger per section)	0.15	m^2
mass flow rate H_2SO_4 (m_S)	0.100	kg/s
mass flow rate Helium per finger (m_H)	0.213	kg/s
pressure drop on shell side	18.9	psi
pressure drop in outer annulus	9.2	psi
pressure drop in inner annulus	1.4	psi

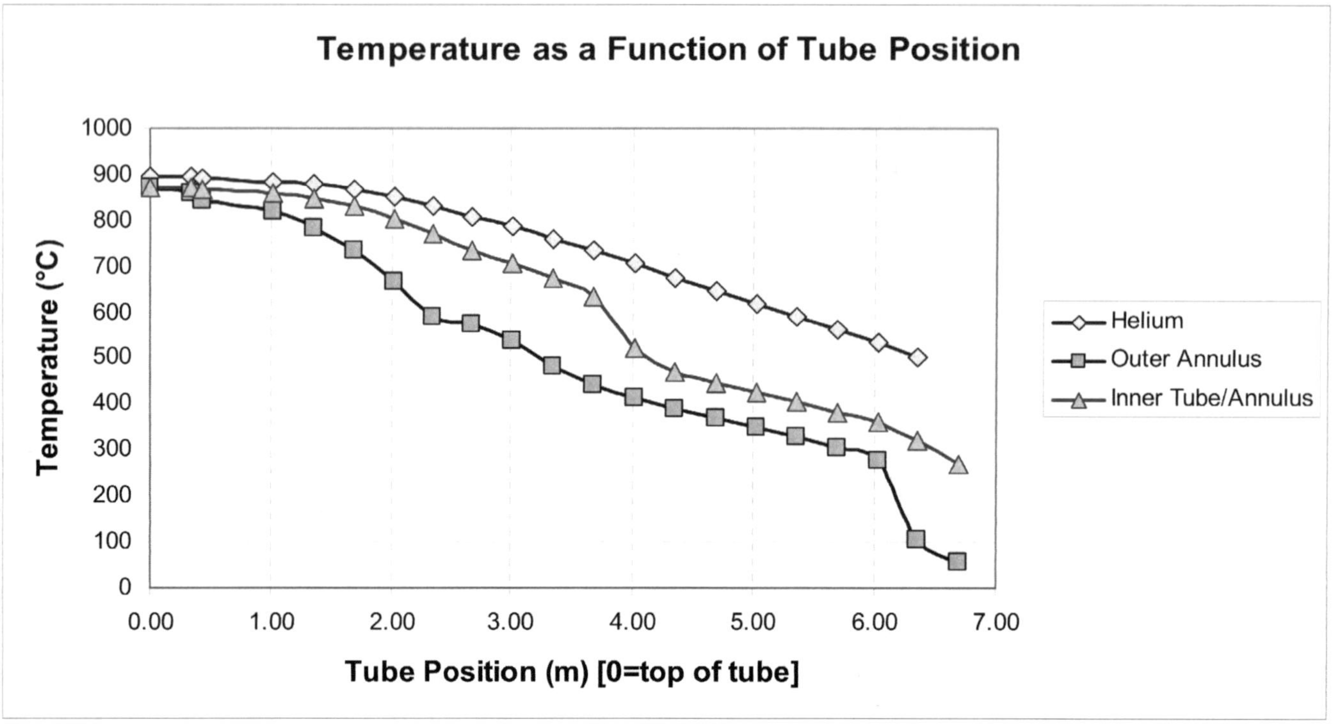

Figure 9 - Temperature Profile for Decomposition Reactor

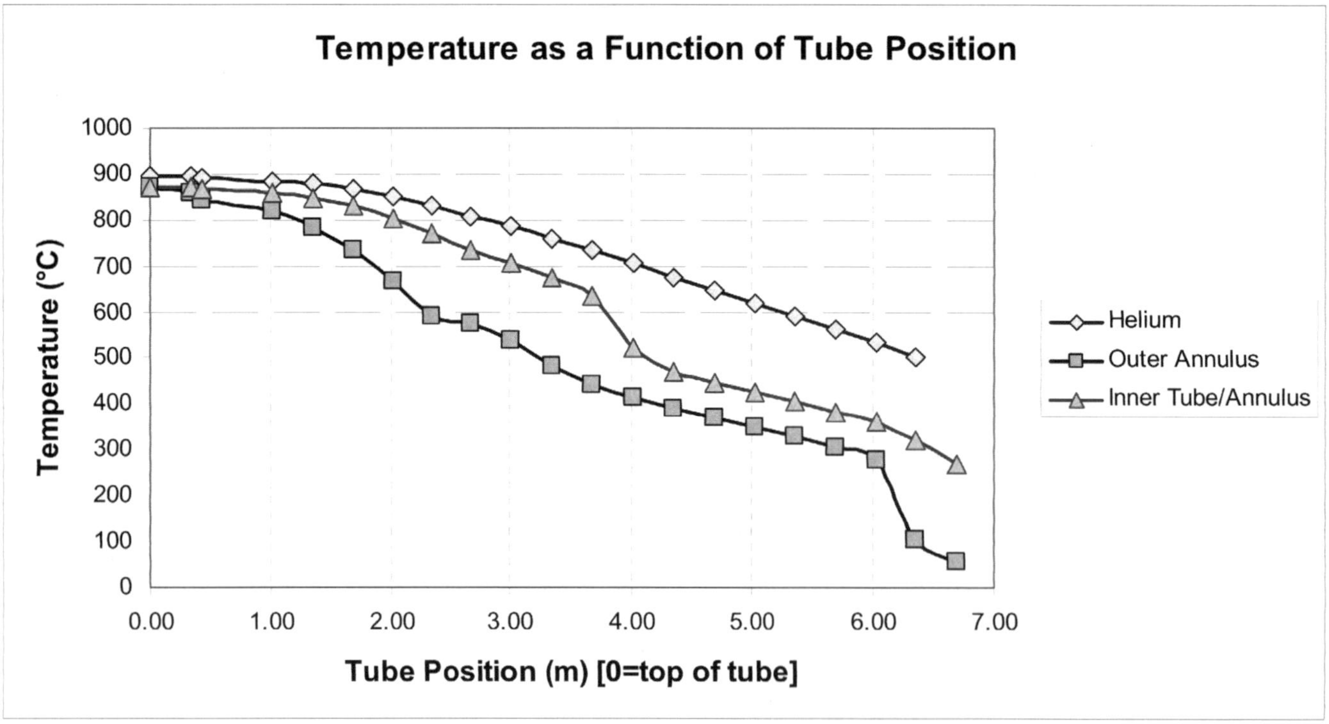

Figure 10 - Pressure Drop for Decomposition Reactor

5. Calculation of Vessel Size and Catalyst Volume

5.1. Vessel Sizing and Concentrator Height

The calculation of the inner diameter of the vessel is based on the total number of bayonet tubes as determined by the decomposition model. The tubes will have a triangular pitch arrangement with a minimum pitch-to-diameter ratio of 1.125. The user can input a desired number of vessels to obtain the required vessel diameter as shown in Table 2.

Table 2 - Calculation of Vessel Diameter

Area of Unit Cell	0.0139	m^2
Area of Tube	0.0082	m^2
Total Cross-Sectional Area	20.92	m^2
Inside Diameter of Vessel	3.65	m
Number of Vessels	2	

5.2. Calculation of Required Catalyst Volume

The required catalyst volume to decompose the SO_3 into SO_2 was calculated to ensure that the annular space in the bayonet tubes were large enough to accommodate the catalyst. The calculation was based on kinetic data from Lin and Flaherty[(6)]. The catalyst was assumed to be iron oxide. Assuming an SO_3 conversion of 50%, the required catalyst volume space is approximately 0.005 m^3 and the available volume in the annular region of the bayonet tubes where the temperature is above 600°C is 0.010 m^3. Thus, the volume in the annular region exceeds the required volume by an order of magnitude and achieving the necessary residence time should not pose an issue.

Appendix A: Pressure Drop Calculations

i. Packed Bed Flow[(3)]

$$\frac{\Delta P}{\Delta Z} = \frac{150 \mu v_c}{L_c^{2}} + \frac{75 \rho v_c^{2}}{L_c}$$

ii. Flow through Pipe[(4)]

$$\frac{1}{\sqrt{f_f}} = 4\log_{10}\frac{d_s}{e} + 2.28 - 4\log_{10}\left(4.67\frac{d_s/e}{\mathrm{Re}\sqrt{f_f}} + 1\right)$$

$$\Delta P = 2 f_f \frac{Z}{d_s} \rho v^2$$

iv. Packed Helium side (structured Packing)[5]

$$\frac{\Delta P}{\Delta Z} = \frac{0.0453\rho_g}{S\varepsilon^2(1-\varepsilon)(\sin\theta)^2} + \frac{6.6441\mu_g}{S^2\varepsilon(1-\varepsilon)\sin\theta}$$

Nomenclature

A_1 – area available for heat transfer between the helium side and the bayonet [m^2]
A_2 - area available for heat transfer between bayonet and support tubes [m^2]
c_p – heat capacity [J/kg/°C]
f_f – Fanning friction factor
h - heat transfer coefficient [$W/m^2/°C$]
H_{in} – enthalpy at the inlet temperature [kJ/kg]
H_{out} – enthalpy at the outlet temperature [kJ/kg]
h_s - heat transfer coefficient on shell side [$W/m^2/°C$]
k – thermal conductivity [W/m/°C]
L_c – characteristic length of packing [m]
$\dot{m}_{gas}$ – mass flow rate of vapor phase in concentrator [kg/sec]
$\dot{m}_{He}$ - mass flow rate of the helium [kg/sec]
$\dot{m}_{H2SO4}$ - mass flow rate of the H_2SO_4 [kg/sec]
$\dot{m}_{liq}$ – mass flow rate of liquid phase in concentrator [kg/sec]
$\dot{n}_{vap}$ – molar flow rate of evaporated water [mole/sec]
$Nu_{boiling}$ – Nusselt correlation for boiling phase
Nu_{gas} – Nusselt correlation for gas phase
Nu_{liq}– Nusselt correlation for liquid phase
Pr – Prandtl number = $\mu c_p/k$
psi – pounds/square inch
S – side dimension of structured packing corrugation [m]
T_{inHe} – inlet temperature to the helium side [°C]
T_{inIA} – inlet temperature to the inner annulus [°C]
T_{inOA} - inlet temperature from the outer annulus [°C]
T_{outHe} – outlet temperature on the helium side [°C]
T_{outIA} – outlet temperature from the inner annulus [°C]
T_{outOA} - outlet temperature from the outer annulus [°C]
$U_{He/OA}$ – overall heat transfer coefficient between the helium and the outer annulus [$W/m^2/°C$]
$U_{OA/IA}$ – overall heat transfer coefficient between the outer and inner annuli [$W/m^2/°C$]
v_c - characteristic velocity [m/s]
VF – vapor fraction
v_s - superficial velocity [m/s]
Z – height [m]

ε – void fraction of packing
μ_g - viscosity of gas [kg/m/s]
ΔH_D – change in enthalpy in the bayonet tube [kJ/kg]
ΔH_{gas} – change in enthalpy of vapor phase in concentrator [kJ/kg]
ΔH_{He} – change in enthalpy of the helium [kJ/kg]

ΔH_{liq} – change in enthalpy of liquid phase in concentrator [kJ/kg]
ΔH_S – change in enthalpy in the support tube [kJ/kg]
ΔH_{vap} – heat of vaporization of water [J/mol]
ΔT_{LM} – log mean temperature difference = $[(T_{inhot} - T_{outcold}) - (T_{outhot} - T_{incold})] / \ln [(T_{inhot} - T_{outcold}) / (T_{outhot} - T_{incold})]$ [°C or K]
$\Delta T_{LMHe/OA}$ - log mean temperature difference between the helium side and inner annulus
$\Delta T_{LMOA/IA}$ – log mean temperature difference between the outer and inner annuli
$\Delta P/\Delta Z$ – pressure drop per unit height [Pa/m]
Φ - ratio of viscosity in the bulk fluid to viscosity at the wall
θ – angle with the horizontal for corrugation channel [radians]
ρ_L- density of liquid [kg/m^3]
ρ_g - density of vapor [kg/m^3]
μ – viscosity [kg/m/s]
ψ – density of water/density of liquid

References:

1. Giazitzoglou, Z. and Wüster, G., *"Das binäre System H_2SO_4; p,v,T-Daten und Dampfdruckkurven in weiten Konzentrationsbereichen bei erhöhtem Druck"*, Ber Bunsenges. Phys. Chem., 85: 127-132, 1981.

2. Connolly, S.M., Zabolotny, E., McLaughlin, D.F., and Lahoda, E.J. "*Design of Composite Sulfuric Acid Decomposition Reactor, Concentrator and Preheater for Hydrogen Generation Process*", Westinghouse, 2007.

3. Mills, Anthony F. Heat Transfer. Irwin, Inc., 1992, p 324-325.

4. Welty, James R. et al. Fundamentals of Momentum, Heat, and Mass Transfer, 4th Ed. John Wiley & Sons, Inc., 2001.

5. Uresti-Melendez, Juana and J. Antonio Rocha. "*Pressure Drop in Ceramic Structured Packings"*, Ind. Eng. Chem. Res. (32), 1992, 2247-2253.

6. Lin, S.S. and R. Flaherty. *"Design Studies of the Sulfur Trioxide Decomposition Reactor for the Sulfur Cycle Hydrogen Production Process"*, Int. J. Hydrogen Energy (8), 32: 589-596, 1983.

391a Solar Cadmium Hydrogen Cycle Update

Bunsen Y. Wong[1], Lloyd C. Brown[1], Robert T. Buckingham[1], and Roger Rennels[2]. (1) General Atomics, 3550 General Atomics Ct., San Diego, CA 92121, (2) Mechanical Engineering, University of Nevada at Las Vegas, 4505 Maryland Parkway, Box 454027, Las Vegas, NV 89154-4027

Flowsheet evaluation of the cadmium solar thermochemical hydrogen generating cycles has indicated thermal efficiencies of nearly 70% (58% lhv) and this has the potential of supplying low cost hydrogen made with a renewal energy source. Laboratory demonstration of the thermodynamics and kinetics of the associated chemical reactions has been on going to validate the possible application of using such cycles on a solar tower. Dissociation experiments of cadmium oxide under various gas compositions have been conducted in order to determine the optimal dissociation process setting. The recombination rates of post decomposition cadmium and oxygen vapors at various temperatures have been measured. This provides input to the design of a quenching step to maximize the cadmium yield. Finally, hydrolysis using molten cadmium at different temperatures has been carried out and results from a rotary kiln designed to demonstrate the hydrolysis process for an on sun demonstration will be presented. An update flowsheet incorporating the new processing information will also be introduced.

391b Aspen Plus Modeling of the Three-Reaction Version of the Copper-Chloride Thermochemical Cycle for Hydrogen Production from Water

Alexandra I. Lupulescu, **John C. Prindle**, and Victor J. Law. Chemical & Biomolecular Engineering, Tulane University, 300 Boggs Building, New Orleans, LA 70118

Aspen Plus Modeling of the Three-Reaction Version of the Copper-Chloride Thermochemical Cycle for Hydrogen Production from Water

The Copper-Chloride thermochemical cycle for producing hydrogen from water has the attractive feature of low temperature requirements. A further desirable property is that possibly the cycle can be implemented such that minimal solids transfers are required. Previous modeling efforts for this cycle involved a four-reaction scheme as follows:

$4CuCl \rightarrow 2CuCl_2(a) + 2Cu$ (electrochemical) 25°C

$2Cu + 2HCl(g) \rightarrow 2\ CuCl(l) + H_2(g)$ 450°C

$2CuCl_2 + H_2O(g) \leftrightarrow Cu_2OCl_2 + 2HCl(g)$ 375°C

$Cu_2OCl_2 \rightarrow 2CuCl + O_2(g)$ 550°C

The most serious drawback of the four-reaction cycle is the presence of elemental copper. Therefore, a three-reaction scheme has been proposed as follows:

$2CuCl + 2HCl \rightarrow 2CuCl_2 + H_2$ (electrochemical) 100°C

$2CuCl_2 + H_2O \leftrightarrow Cu_2OCl_2 + 2HCl$ (vacuum) 375°C

$Cu_2OCl_2 \rightarrow 2CuCl + 0.5O_2$ 550°C

To our knowledge, no prior work has been done on mechanistic modeling of the electrolyzer for the hydrogen production step of the three-reaction scheme. Prior simulations of the four-reaction step used a simple stoichiometric reactor model for the electrolyzer. In the current work, we present an Aspen Plus model for the three-reaction system that includes a realistic model (written as a user-supplied Fortran model) of the electrolyzer.

We developed the Aspen Plus model in three steps. First, a flowsheet was developed in which all reactors are stoichiometric with reactions going to completion. The second phase included the $CuCl_2$ hydrolysis step as an equilibrium reactor. The final step involves the inclusion of the Fortran code to describe the operation of the electrolyzer.

The currently proposed Aspen flowsheet with the actual process implementations described in this paper has two significant advantages:

- No Compressors (Only Pumps & Throttling Valves)
- No Solids Transfer Steps (With Fluidized Bed for O_2/HCl Production Section)

Other topics that are addressed in the paper include the following:

- Continued Overall Process Flowsheet Development
- Sensitivity of Electrolyzer Parameters to Process Conditions
- Water/HCl Purification
- Materials of Construction

391c High Temperature Rapid Dissociation of Mn_2O_3 In An Aerosol Flow Reactor

Todd M. Francis, Christopher Perkins, Paul Lichty, and Alan W. Weimer. Department of Chemical and Biological Engineering, University of Colorado, 1111 Engineering Drive, Boulder, CO 80309-0424

A three-step metal oxide cycle to produce renewable hydrogen has been proposed. This cycle employs manganese oxide and solar energy to produce hydrogen:

$\frac{1}{2}Mn_2O_3 \rightarrow MnO + \frac{1}{4}O_2$ (1)

$MnO + NaOH \rightarrow \frac{1}{2}H_2 + NaMnO_2$ (2)

$NaMnO_2 + \frac{1}{2}H_2O \rightarrow \frac{1}{2}Mn_2O_3 + NaOH$ (3)

$H_2O \rightarrow H_2 + \frac{1}{2}O_2$ (4)

Most work on the manganese oxide cycle has focused on the hydrogen generating step (Equation 2) and the product recovery step (Equation 3) with little work done on the reduction step (Equation 1). It is essential to understand the reduction step because the entire feasibility of the manganese oxide cycle is based on this reaction having a high overall conversion. Specifically, the viability of the MnO and O_2 phase separation must be investigated to determine how much recombination might occur during the quench step. In this study, experimental designs were done to investigate the affect of oxygen on the conversion of Mn_2O_3 to MnO. Along with the results from this study, results from a CFD simulation of Mn_2O_3 through an aerosol flow reactor will be presented.

391d Size-Dependent Zn Nanoparticle Hydrolysis for Hydrogen Generation

X. Ma[1], Anshuman Lall[2], G. M. Mulholland[2], and Michael Zachariah[3]. (1) University of Maryland-College Park, College Park, MD 20742, (2) Department of Mechanical Engineering, University of Maryland-College Park, College Park, MD 20742, (3) University of Maryland and NIST, 2125 Glenn L Martin Hall, College Park, MD 20742

A novel tandem ion mobility method is developed to study the size-dependent kinetics of Zn nanoparticle hydrolysis reaction, which is the second, exothermic step of Zn/ZnO Two-Step Water-Splitting Cycle for hydrogen generation. The experiment consists of two different ion-mobility schemes in series. The first mobility characterization is to size select particles with a differential mobility analyzer (DMA). The second mobility characterization employs an aerosol particle mass analyzer (APM) and measures changes in mass resulting from a controlled hydrolysis reaction of the Zn nanoparticles. The intrinsic reaction rates are determined for particles of different initial mobility sizes. The unsupported single Zn nanoparticles are generated using a condensation-evaporation method. It is found that smaller particles have a higher reaction rate which favors hydrogen generation. Electron microscopic analysis is employed to study the morphology change of oxidized Zn nanoparticles.

The same tandem ion mobility method is also used to study the reaction kinetics for Zn nanoparticles oxidized in air at elevated temperatures. Compared with the hydrolysis reaction, the air oxidation kinetics are much faster while the onset temperature of reaction is higher than hydrolysis reaction for the same starting particle size. The latter may be due to the effect of fast water ionization reaction which happens at the surfaces of Zn nanoparticles.

This paper will give quantitative rates for hydrolysis and oxidation as a function of particle size.

391e Kinetics of the Thermochemical Ferrite Water Splitting Cycle Using Nano-Thick Cobalt-Iron Spinel Oxide Films Synthesized Via Atomic Layer Deposition

Jonathan R. Scheffe[1], Nathan P. Siegel[2], Mark D. Allendorf[2], and Alan W. Weimer[1]. (1) Department of Chemical and Biological Engineering, University of Colorado, 1111 Engineering Dr., ECCH 111, Boulder, CO 80309-0424, (2) Solar Technologies, Sandia National Laboratories, PO Box 5800, Albuquerque, NM 87185

Nano-thick cobalt-iron spinel oxides (cobalt ferrites) have been synthesized via atomic layer deposition (ALD) on nano-scale yttria-stabilized zirconia (YSZ) and Al_2O_3 supports. ALD has the advantage of being able to precisely control the thickness and Co/Fe ratio on the atomic level, which is significant because it has been suggested that this two-step cycle is a surface dominated reaction. The kinetics of the water splitting step has been investigated and results indicate that the temperature required is decreased due to a reduction of diffusion limitations. Additionally, results were directly compared to samples synthesized via traditional methods (coprecipitation, etc), and the amount of hydrogen generated per mole of ferrite was shown to be greater. Potential factors affecting the water splitting step that were studied included film thickness, Co/Fe ratio, and the substrate material (YSZ and Al_2O_3), and each was shown to be statistically significant. Hydrogen evolution was measured in situ using a residual gas analyzer, and changes in crystallinity were measured using a powder x-ray diffractometer.

455a Hybrid Sulfur Cycle Flowsheets for PEM Electrolyzers with SO2 Vapor Feed

Maximilian B. Gorensek, Computational Science and Statistics Department, Savannah River National Laboratory, Aiken, SC 29808 and John W. Weidner, Swearingen Engineering Center, Chemical Engineering Department, University of South Carolina, 301, S Main Street, Columbia, SC 29208.

Hybrid Sulfur cycle flowsheets that have been developed for the DOE Nuclear Hydrogen Initiative typically feature SO2-depolarized electrolyzers (SDEs) in which the anode is fed a solution of SO2 dissolved in sulfuric acid. An alternative SDE configuration, in which the anode is fed wet SO2 vapor, is also being considered. This offers several advantages as well as disadvanates when compared to the liquid-fed version. Changes in the Hybrid Sulfur flowsheet necessitated by switching to an SO2 vapor-fed SDE are examined in this paper. Flowsheet performance is estimated based on the results of Aspen Plus simulations and compared with that of earlier flowsheets that use liquid-fed SDEs.

455b Conceptual Design and Cost Estimate for a Solar-Powered Hybrid Sulfur Hydrogen Production Process

William A. Summers[1], Maximilian B. Gorensek[2], and Claudio Corgnale[1]. (1) Hydrogen Technology Center, Savannah River National Laboratory, Savannah River Site, 773-42A/137, Aiken, SC 29808, (2) Computational Science and Statistics Department, Savannah River National Laboratory, Aiken, SC 29808

This paper will describe a solar-powered Hybrid Sulfur cycle thermochemical hydrogen production process conceptual design developed at Savannah River National Laboratory. The design leverages Hybrid Sulfur cycle process developments made under the Nuclear Hydrogen Initiative. An H2A analysis based on Aspen Plus flowsheet simulation results was used to estimate production costs.

455c Generation of Hydrogen Using a Multicell Stack Electrolyzer with Sulfur Dioxide Depolarized Anode

John L. Steimke, Environmental and Chemical Process Technology, Savannah River National Laboratory, Building 786-5A, Savannah River Site, Aiken, SC 29808 and David T. Herman, Separations Science, Savannah River National Laboratory, Building 735-11A, Savannah River Site, Aiken, SC 29808.

The DOE Office of Nuclear Energy, Science & Technology has established the Nuclear Hydrogen Initiative (NHI) to develop technologies that can be coupled with next generation nuclear reactors for hydrogen production. One process under development, the Hybrid Sulfur (HyS) Process, generates hydrogen using electrolyzers with sulfur dioxide depolarized anodes. The Savannah River National Laboratory (SRNL) first received funding to test and develop this type of electrolyzer in 2005 and received funding for continued testing through FY 08.

Traditional water electrolyzers have an anode and cathode separated by an electrically insulating membrane that does not pass product gases. Water is fed to the anode where it is oxidized to oxygen and hydrogen ions. The ions cross the membrane where they are reduced to hydrogen gas. The required voltage is between 1.5 volts and 2 volts. For a sulfur dioxide depolarized electrolyzer, the reactants water and sulfur dioxide are fed to the anode where they react to form sulfuric acid and hydrogen ions (protons). Similarly to the traditional electrolyzer, the hydrogen ions cross the membrane and are reduced to hydrogen. The cell potential can be significantly less than one volt, which is important because electrolyzer power cost is proportional to cell potential. A separate high temperature reaction decomposes sulfuric acid to regenerate sulfur dioxide.

The SRNL electrolyzer system operates by feeding SO2 dissolved in sulfuric acid to the anode. The anode of the electrolyzer is bathed in a sulfuric acid solution saturated in sulfur dioxide. Sulfur dioxide is oxidized to sulfuric acid at the anode. Hydrogen ions cross a membrane to the cathode where they are reduced to hydrogen gas. The cathode is flushed with water for temperature control and to sweep away any material that crosses the membrane.

SRNL designed, constructed and tested a facility built around a single cell sulfur dioxide depolarized electrolyzer. As part of the continued development of the HyS program, the test facility was modified to accept a three cell stack electrolyzer, expanding the active cell area from 60 cm2 480 cm2. The facility allows pressures as high as 75 psig and temperatures as high as 80ºC. Modifications to the facility included increasing the power input, adding additional flow lengths to provide current isolation for each cell, and the ability to preheat the cathode water flush.

The Three Cell Stack was designed jointly by Giner Compay and SRNL and was constructed by Giner Co. The stack has three cells and a total area of 480 cm2. The three cells are connected electrically in series and hydraulically in parallel. A challenge for the stack was to limit leakage currents through the conductive fluid in the anolyte flow loop. Testing includes two different membrane electrode assembly (MEA) configurations and under different operating conditions. The maximum hydrogen production rate was 80 standard liters per hour. Individual cell voltages were measured and ranged between 0.61 and 1.2 volts with current densities up to 400 mA/cm2. After completing tests with the first set of MEAs, the stack was removed from the test stand, disassembled and inspected. The MEAs were removed and analyzed by Scanning Electron Microscopy (SEM) for the potential deposition of elemental sulfur. The electrolyzer was then re-assembled using a second set of MEAs with different catalyst loadings and testing continued.

455e Westinghouse Cycle Electrochemical Step: Development Status and Modeling at Cea-Marcoule Laboratory

Sophie Charton[1], Denis Ode[1], Patrick Rivalier[1], Janvier Josselin[1], Jean Pierre Caire[2], and Eric Chainet[2]. (1) Fuel Cycle Department Technology, Commissariat à l'Energie Atomique, Centre de Marcoule - BP17171, Bagnols Sur Ceze, 30200, France, (2) ENSEEG - INP Grenoble, 1130 chemin de la piscine, Saint Martin D'heres, 38400, France

Among the thermo chemical cycles conceivable for hydrogen massive production coupled with a nuclear power source, two are currently considered and compared at the Nuclear Energy Division of the French CEA: the iodine-sulfur cycle (based on the Bunsen reaction) and the hybrid sulfur cycle based on depolarized SO2 oxidation (the Westinghouse cycle).

While prospects on the iodine-sulfur cycle has led to the realization of a joint SNL-GA-CEA test plant, which is today operated in San Diego [ref1]; our studies regarding the Westinghouse process were mainly theoretical and prospective.

In this scope, the electrochemical step has been modeled considering a filter-press technology; and, although few literature data were available, contributions of the major electrochemical and fluid mechanics factors to the overall cell voltage have been analyzed by numerical simulation [ref2].

Despite their different degrees of achievement, the investigations led on both cycles within this exploratory phase have enabled their respective key issues identification and the evaluation of their energy requirements. The hybrid cycle relevance was emphasized.

This assessment is nevertheless based on generic materials and technologies. A test plant dedicated to electrolysis cells operating has therefore been designed and is running in Marcoule Laboratory since April'08 [ref3].

Thanks to this versatile tool, hydrogen production resulting from depolarized SO2 oxidation is currently studied under a wide range of operating conditions (electrolytes flow rates, concentrations and temperatures). Cell voltages measured under galvanostatic conditions are analyzed and compared with model predictions, thus allowing a better understanding of the observed overall performances.

At last, model development status and needs are discussed. Indeed, a more detailed description of i) the electrochemical process, ii) species transport and membrane exchange, iii) phases equilibriums, iv) hydrogen bubbles behavior, is required in order to reach a sufficient degree of understanding of the electrolytic cell functioning, and therefore achieving the design and optimization of an industrial-scale process.

References:

[ref1] J. Leybros et al., GLOBAL Meeting, Boise, Sept. 2007.

[ref2] F. Jomard et al., Journal of Applied Electrochemistry, 38, 2008, pp.297.

[ref3] P. Rivalier et al., 16th ICONE Meeting, Orlando, May 2008.

599a Dynamic Simulation of a Coupled Heat Transfer Loop and a Nuclear Reactor In a Nuclear Hydrogen Plant

Patricio D. Ramirez-Munoz[1], Mujid S. Kazimi[2], and Paul I. Barton[1]. (1) Chemical Engineering, Massachusetts Institute of Technology, 77 Massachusetts Ave., RM 66-363, Cambridge, MA 02139, (2) Center for Advanced Nuclear Energy Systems, Department of Nuclear Science and Engineering, Massachusetts Institute of Technology, 77 Massachusetts Ave., RM 24-219, Cambridge, MA 02139

The study of nuclear hydrogen plant operation and safety is an important next step in assessing the desirability of bringing this technology to reality. The overall system for nuclear hydrogen production consists of the nuclear plant, the power conversion unit (PCU), the heat transfer loop and the hydrogen plant. This system has been studied using steady-state models and has been shown to have the potential to achieve high efficiencies for hydrogen production [8]. Steady-state models have allowed optimization of the plant configuration and operating conditions [1, 2, 4, 6, 7]. The dynamic interaction of all the components of the system should be investigated next, to produce a more robust and stable system, avoiding accidents and improving operability. Questions regarding maximum stresses exerted on materials during transients and how to modulate the heat delivered to the chemical plant in a safe manner must be addressed.

Nuclear power and hydrogen production systems have distinct dynamic characteristics and their interaction can produce results hard to predict intuitively. Therefore, it is necessary to use mathematical models describing the transient behavior of these systems. Dynamic models have assisted engineers for several decades in the assessment of safety of nuclear power plants [3, 9]. On the other hand, hydrogen production alternatives utilizing heat and/or electricity from nuclear power plants have been mostly studied within the steady-state regime to make design decisions. Some efforts have been made to understand the dynamics of this system [5, 10]. The hydrogen production plant adds extra complexity to the system because it includes chemical reactions with nonlinear dynamics. The simulation environment used should be able to capture these physical and chemical phenomena. The long term goal of this research is to develop a suite of mathematical models that enable dynamic simulation, sensitivity analysis and optimization studies of the interaction of a nuclear reactor and a hydrogen generation plant.

This paper focuses on dynamic simulation of a coupled system comprising a nuclear reactor and a heat transfer loop. The nuclear reactor generates 600MWth and part of this heat is used to generate high-temperature steam to feed the hydrogen plant. High-temperature electrolysis is considered as the electrochemical process to produce hydrogen. The system has two loops and its components are a nuclear reactor, two heat exchangers, piping and two compressors. The nuclear reactor is a Pebble Bed Modular Reactor (PBMR) and the heat exchanger removing the heat from the nuclear reactor coolant is a compact heat exchanger. The heat exchanger heating the steam is of the tube-in-shell kind [1] and each compressor regulates the gas flowrate in each loop. Helium is considered as the heat transfer fluid for both loops.

The system model encompasses models for each one of the units mentioned previously. The dynamics of the helium in the heat transfer loop were represented with a simplification of the inviscid Navier-Stokes equations [7]. The system of equations was discretized in the axial coordinate in order to calculate detailed state variable profiles around the loop. The nuclear reactor was represented with a model developed by Chunyun Wang at MIT [11], with coupled point-kinetics models, reactivity models and thermal-hydraulic models. The heat exchangers were represented by discretized energy balances. This system model was implemented and solved using JACOBIAN®, a state-of-the-art dynamic simulation software. JACOBIAN® uses an implicit integrator and it can handle differential-algebraic equations and discrete/continuous models.

A simple start-up scenario was used to study the behavior of the system. The start-up schedule had four stages, and the first one was to stabilize the system at a low power level. This was achieved by running the compressors and nuclear reactor at 1% of the designed power level and waiting until the system reached a steady state. Second, the power in both compressors was increased to boost the heat removal capacity of the system, and the flow of steam on the process heat exchanger (PHX) was also increased. Third, the heat generated in the nuclear reactor was boosted until it reached the design point. Finally, the temperature in the cold side of the PHX was raised, representing the start-up of the chemical plant.

1. C. B. Davis, R. B. Barner, S. R. Sherman, D. F. Wilson. Thermal-Hydraulic Analyses of Heat Transfer Fluid Requirements and Characteristics for Coupling a Hydrogen Product Plant to a High-Temperature Nuclear Reactor. Technical Report INL/EXT-05-00453, Idaho National Laboratory, 2005.

2. Y. H. Jeong, K. Hohnholt, M. S. Kazimi and B. Yildiz. Optimization of the hybrid sulfur cycle for hydrogen generation. Technical Report MIT-NES-TR-004, Center for Advanced Nuclear Energy Systems, Massachusetts Institute of Technology, 2005.

3. Michitsugu Mori. Simulation of a BWR-5 plant: Verification and validation of RETRAN-03. Nuclear Technology, 121:245-259, 1998.

4. C. H. Oh, R. Barner, C. B. Davis and S. R. Sherman. Evaluation of working fluids in an indirect combined cycle in a very high temperature gas-cooled reactor. Nuclear Technology, 156:1-10, 2006.

5. S. Oh, N. Brown and S. T. Revankar. Transient Model for the Chemical Chamber in Sulfur Iodine Thermo-Chemical Process, Technical Report PU/NE-06-010, School of Nuclear Engineering, Purdue University.

6. S. B. Rodríguez Jr., R. O. Gauntt, S. T. Revankar and K. Vierow. Development of design and simulation model and safety study of large-scale hydrogen production using nuclear power. 2005 AIChE Annual Meeting and Fall Showcase, Conference Proceedings, 2005, 8018-8023.

7. Todreas N. E. and Kazimi M.S. Elements of thermal hydraulic design. Hemisphere Publishing Corporation, New York, 1990.

8. B. Yildiz, K. Hohnholdt, and M. S. Kazimi. Hydrogen production using high temperature steam electrolysis supported by advanced gas reactors with supercritical CO2 cycles. Technical Report MIT-NES-TR-002, Center for Advanced Nuclear Energy Systems, Massachusetts Institute of Technology, 2004.

9. B. Yildiz and M. S. Kazimi. Efficiency of hydrogen production systems using alternative nuclear energy technologies. International Journal of Hydrogen Energy, 31:77-92, 2006.

10. K. Vierow, Y. Liao, J. Johnson, M. Kenton, and R. Gauntt. Severe accident analysis of a PWR station blackout with the MELCOR, MAAP4 and SCDAP/RELAP5 codes. Nuclear Engineering and Design, 234:129-145, 2004.

11. R. B. Vilim. Dynamic Modeling Efforts for System Interface Studies for Nuclear Hydrogen Production. Argonne National Laboratory, ANL-07/16, 2007.

12. C. Wang. Design, Analysis and Optimization of the Power Conversion System for the Modular Pebble Bed Reactor System. PhD thesis, Massachusetts Institute of Technology, 2003.

599b Materials Characterization for High Temperature Solar Hydrogen Production

Paul Lichty[1], Theres Felix[2], Amanda Scott[3], and Alan Weimer[1]. (1) Department of Chemical and Biological Engineering, University of Colorado, 1111 Engineering Drive, Boulder, CO 80309-0424, (2) ETH, Zurich, Switzerland, (3) Chemical Engineering, Vanderbilt University, 2301 Vanderbilt Place, VU Station B #4003, Nashville, TN 37235

Concentrated solar energy focuses large amounts of sunlight onto a small area producing high temperatures (~2000°C). A novel multi-tube solar thermal aerosol reactor has been designed for testing at the High Flux Solar Furnace(HFSF) at the National Renewable Energy Laboratory(NREL) in Golden, Colorado. The reactor consists of a reflective outer cavity which houses several absorbing, reaction containing tubes. Concentrated solar energy impinges upon the absorbing tubes through a window in the outer chamber. Ray trace modeling of the HFSF coupled with experimental characterization has provided power and flux profile boundary conditions entering the chamber. Usable reactor materials must be capablc of withstanding high temperature corrosive environments and rapid temperature variations. Rapid changes in temperatures cause internal material stresses. An investigation of the typical stresses produced through high temperature thermal cycling at the HFSF provides criteria for selecting functional materials. Candidate materials are subjected to thermal cycling. Tensile stress samples undergo rapid heating and cooling at the HFSF. Residual strength is measured through tensile testing of the cycled samples. Stress estimations as well as lifetime estimates are analyzed from the residual strength data. These tests coupled with computational thermal stress analysis will provide a stress history specific to the flux profile produced by the HFSF. Localized stress intensity factors will be calculated from modeling and compared to experimental results. Stress intensity factors will be quantified locally by subjecting large planar samples with multiple, distributed, vicker's indentations to the HFSF flux profile. Analysis of crack propagation as a function of indentation location and number of cycles will provide validation of modeled stress intensity. Preliminary results from these tests will be presented as well as comparison to computational model.

599c Magnesian Calcite Sorbent for Carbon Dioxide Capture during Hydrogen Production from Fossil Fuels

James Mabry[1], Kanchan Mondal[2], and Tomasz Wiltowski[2]. (1) Department of Mechanicla Engineering and Energy Processes, Southern Illinois University, Carbondale, IL 62901, (2) Department of Mechanical Engineering and Energy Processes, Southern Illinois University, Carbondale, IL 62901

A magnesian calcite type material was synthesized for the removal of carbon dioxide. The results from characterization, reactivity and CO2 capture capacity for different synthesis conditions are reported. In the synthesis process a magnesian calcite material CaCO3:MgCO3 is synthesized by the co-precipitation of specific amounts of commercially available CaO and MgO by carbon dioxide. Characterization was done with BET, SEM/EDS, particle size analysis, and XRD. The reactivity and capacity was measured using TGA cycles at 750, 800 and 850°C, and long term tests (50 cycles) was run on the optimal sample. Reaction cycles were carried out under 100% carbon dioxide and the regeneration cycles were run under nitrogen. The effects of CaO, MgO, and surfactant loading were studied to determine the maximum amount of carbon dioxide removal for the sorbents and correlated to particle size and pore distribution of the sorbents. The kinetics of the CO2 capture process in commercially available limestone and dolomite and those with laboratory synthesized calcium oxide and laboratory synthesized magnesian calcites are also presented.

599d On the Design of Hydrogen Storage Vessels Utilizing a Complex Hydride Storage Material

Yongfeng Wang, Armin D. Ebner, and James A. Ritter. Department of Chemical Engineering, University of South Carolina, Columbia, SC 29208

Hydrogen is being considered very seriously as an alternative, renewable energy carrier for use in internal combustion engines or fuel cells [Ritter et al, 2003]. However, despite some recent advancement, hydrogen storage is still a major concern. In addition to cost, safety and achieving sufficient gravimetric and volumetric densities, a hydrogen storage system, for both stationary and mobile applications, requires suitable thermal integration to facilitate optimal charging and discharging.

Among the commercially viable solid matrix methods of storing hydrogen, the complex hydride, NaAlH4, when doped with Ti, is one of the few, perhaps even the best, option for both stationary and mobile applications. The excitement about this system has been due to its relatively high gravimetric density and its reversibility at relatively low temperatures. However, a paucity of information is available on the design and development of hydrogen storage vessels containing complex hydrides.

The primary objective of this work is to further develop an existing 2-D mathematical model [Gadre et al, 2005; Wang et al, 2007a,b] to describe the dynamic behavior of a complex hydride hydrogen storage vessel during uncontrolled charge operations. This objective will be facilitated by drawing on the previous work by the authors on the modeling of more conventional metal hydride hydrogen storage vessels [Gadre et al., 2003; Jiang et al, 2005; Gadre et al, 2005; Wang et al, 2007a,b]. A secondary objective of this work is to study different ways of thermally integrating the hydrogen storage vessel, ranging from internal water or air heat exchangers, to external water or air heat exchangers. Again, this objective will be facilitated by drawing on the previous work by the authors on internal water heat exchanger vessel designs for metal hydride hydrogen storage vessels [Gadre et al., 2003; Jiang et al, 2005; Gadre et al, 2005; Wang et al, 2007a,b].

It is the contention here that the models developed by Ritter and co-workers for metal hydrides can be easily modified to explore the use of Ti-doped NaAlH4 as the reversible hydrogen storage media. Physical, thermodynamic and kinetic properties of Ti-doped NaAlH4, typically as a function of temperature and pressure, are needed. Some of this information is available in the literature, such as conductivity and heat capacities [Lesch et al., 2004; Wang et al., 2004], hydrogen storage capacities [Gross et al., 2002], charge and discharge kinetics [Bellosta von Colbe et al., 2005]. Some of it might have to be estimated. It is also the contention here that these models can be easily modified to explore different heat transfer modes. Different geometries for internal and external fluid heat exchangers will be readily programmed into the Comsol Physics software package, and different heat transfer coefficient correlations will be examined for the internal and external air heat exchanger configurations, such as finned geometries. This presentation will provide an up to date account of this model development effort being carried out for the design of a complex hydride hydrogen storage vessel.

Bellosta von Colbe, J. M., M. Felderhoff, B. Bogdanovic, F. Schueth, C. Weidenthaler, "One-Step Direct Synthesis of a Ti-Doped Sodium Alanate Hydrogen Storage Material, Chemical Communications, 37, 4732-4734 (2005).

Gadre, S. A., A. D. Ebner, S. A. Al-Muhtaseb and J. A. Ritter, "Practical Modeling of Metal Hydride Hydrogen Storage Systems," Ind. Eng. Chem. Res., 42, 1713-1722 (2003).

Gadre S.A., A.D. Ebner and J.A. Ritter, "Two Dimensional Model for the Design of Metal Hydride Hydrogen Storage Systems," Adsorption, 11, 871-876 (2005).

Gross, K.J., G.J. Thomas and C.M. Jensen, "Catalyzed Alanates for Hydrogen Storage", J. Alloys Compd. 683-690 (2002)

Lesch, D A., Lewis G.J., Low, J.J., Sachtler, J.W.A., "Discovery of Novel Complex Metal Hydrides for Hydrogen Storage through Molecular Modeling and Combinatorial Methods" DOE FY 2004 progress report III.C.4

Jiang, Z., R.A. Dougal, S. Liu, S.A. Gadre, A.D. Ebner and J.A. Ritter, "Simulation of a Thermally-Coupled Metal Hydride Hydrogen Storage and Fuel Cell System," J. Power Sources, 142, 92-102 (2005).

Ritter, J. A., A. D. Ebner, J. Wang and R. Zidan, R., "Implementing a Hydrogen Economy," Materials Today, 9, 18-23 (2003).

Wang, J. et al., "Hydride Development for Hydrogen Storage", DOE FY 2004 progress report III.C.2

Wang Y., A.D. Ebner and J.A. Ritter, "Thermal Characterization of a Metal Hydride Hydrogen Storage System with a Two Dimensional Model during Controlled Discharge," Int. J. Hydro. Energy, to be submitted (2007a)

Wang Y., A.D. Ebner and J.A. Ritter, "Thermal Characterization of a Metal Hydride Hydrogen Storage System with a Two Dimensional Model during Rapid Charging," in preparation (2007b).

HIGH TEMPERATURE HEAT EXCHANGER AND COMPONENT TEST LABORATORY – A GRADED APPROACH

Steven R. Sherman, Savannah River National Laboratory, Aiken, South Carolina, USA
Yitung Chen, University of Nevada Las Vegas, Las Vegas, Nevada, USA

Introduction

The Next Generation Nuclear Plant (NGNP) Project and the Nuclear Hydrogen Initiative (NHI) are U.S. Department of Energy programs that are responsible for the research, development, and testing of high-temperature gas-cooled nuclear reactor technologies and thermally driven hydrogen production methods. The two programs are related, in that the nuclear reactor will provide high-temperature heat, and the thermally-driven hydrogen plant will consume it. The nuclear reactor is envisioned as a large-scale (500-600 MWth) high-temperature (750-900°C) high-pressure (5-9 MPa) [1] helium-cooled unit that is capable of producing electricity or high-temperature heat. The hydrogen plant is less defined, but a number of methods are currently under investigation by the NHI including High-Temperature Electrolysis [2], the Hybrid Sulfur process[3], and the Sulfur-Iodine process[4]) and lower-temperature alternatives. A high-temperature heat transfer network (HEN) is needed to deliver thermal energy from the nuclear reactor to the hydrogen plant. The combination of a high-temperature gas-cooled nuclear reactor, a HEN, and a downstream user of high-temperature heat such as a nuclear hydrogen plant is known as the NGNP.

Heat exchangers define the HEN, and heat exchanger materials and designs must be identified that operate efficiently and can withstand high operating temperatures and pressures without suffering creep failure. Above 700°C, even high-temperature alloys such as Inconel 617 suffer reduced yield stress and time-to-rupture due to creep. Therefore, effective heat exchanger designs must be developed that minimize stresses, so that failure due to creep can be avoided. In the longer term, substituting ceramics for metals may be a solution, since ceramics have less tendency to creep at higher temperatures, but ceramic heat exchangers are still relatively immature in comparison to metal heat exchangers and are not yet available for nuclear reactor applications.

At the present time, there is no readily accessible laboratory in the U.S. for operating at the high temperatures, pressures, and flow conditions needed to test components for the NGNP HEN. Fortunately, the test equipment and methods are not new and the capabilities needed to perform high temperature heat exchanger testing are not unique. A dedicated testing laboratory could be established, with some investment, at almost any university or national laboratory.

Laboratory-scale static and flow test beds are described that would be capable of measuring leakage rates, fluid flow characteristics, and heat transfer performance at the temperatures and pressures of interest. A graded approach to testing is recommended that would minimize costs at initial stages, while allowing for more accurate measurements at later stages. Initially, once-through test beds that use heated air or steam are suggested to test heat exchanger performance, and the data collected from these tests will be used to calibrate heat exchanger models. Then, one or more closed helium loops can be constructed to more closely measure heat exchanger performance under more representative fluid conditions. Fluid similitude will be used to choose and compare fluid conditions. The test beds must be flexible and should be able to accommodate most lab-scale heat exchangers and other components as long as their input and output manifolds conform to defined dimensions.

Pressure/Leak Testing

A heat exchanger test laboratory must have the capability to perform hydrostatic pressure testing and leak testing capabilities. A hydrostatic pressure test is performed in order to verify that the prototype device can withstand differential pressures in excess of expected operating pressures without excessive deformation or failure. A heat exchanger that is expected to operate with a steady pressure differential (either internally or externally) of 15 psig or greater falls under the American Society of Mechanical Engineers (ASME) Boiler and Pressure Vessel Code [5], and the code requires that a hydrostatic test be performed. Such a test would be performed with water or other fluid, and at room temperature. Though the heat exchanger will be expected to operate at elevated temperature, successful completion of a hydrostatic test represents the minimum safety threshold for the component before elevated temperature testing can begin.

Pressure testing should be repeated at the temperatures of interest, though high temperatures will most likely require the use of gaseous fluids to provide the pressure differential, and this will lead to greater energy release at failure. A suggested method for performing this test safely at high temperature is to place the prototype component inside a kiln or furnace within a shielded (but vented) container. The shielded container would provide some protection against shrapnel in case of catastrophic component failure, and the walls of the kiln or furnace would provide back-up protection. The kiln or furnace can also be placed behind a barrier to provide a third layer of protection for the experimenter. The test would be performed by placing the component within the shielded container, heating the component and container within the kiln or furnace until it reaches a steady temperature, and then pressurizing the component for the duration of the test. Figure 1 shows a basic drawing of such a pressure test apparatus.

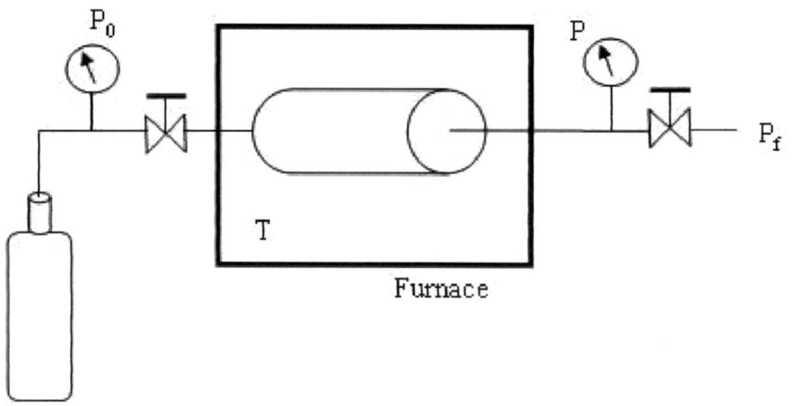

Figure 1: Simplified high-pressure test and leak test apparatus.

Beyond pressure testing, leak testing should be performed at the temperatures and pressure of interest in order to quantify detectable leakage rates from the inside to the outside of the prototype component, and, in the case of heat exchangers, to quantify detectable leakage rates between the hot and cold channels. Leakage rate testing is performed by pressurizing the component at an equilibrium temperature, and then measuring the decay in pressure. A steady-state leakage rate can be deduced from the test conditions and the measured decay rate [6], and the individual leakage rates of a heat exchanger (hot side external, cold side external, cross leakage) can be measured by pressurizing just the hot side of the HX, just the cold side of the HX, and then both sides at the same pressure in order to differentiate between external leakage and cross-heat exchanger leakage [7]. For added accuracy, the leakage tests are performed at different temperatures to more closely match the average temperatures of the hot and cold sides of the heat exchanger.

Beyond leak testing, the proposed leak testing apparatus could be used for performing cyclic pressurization if automatic pressure controls were installed. Such testing is useful if it is desired to determine whether a design or prototype is subject to deleterious effects from mechanical fatigue.

Once-Through Flow Testing

Concerning the NGNP, the most accurate heat exchanger testing would be performed at the conditions and with the fluids of interest. That is, testing would be performed at high-temperature and pressure, and by using helium as the chosen heat transfer fluid. Since Grade-A helium sells for approximately \$3.24 to \$3.79/m^3 (2007 \$) [8], closed heat transfer loops would be needed to conserve helium. Closed loops require the use of helium compressors and other control systems, so that the overall cost of the test bed would be relatively expensive.

The point of initial lab-scale tests, however, is to measure performance data in order to validate heat exchanger models, and exact replication of the fluid conditions of interest is less important. As long as good instrumentation is used and proper testing procedures are adopted, the heat exchanger models can be validated against the tested fluid conditions, assuming that the models are applicable to both helium and non-helium fluids. It is suggested that air and/or steam be used as heat transfer fluids, and that an open flow systems that rely on pressurized sources be used to create the fluid flow conditions of interest rather than compressors. Air and water are freely available in nearly any laboratory at much less cost per unit volume than helium.

Figure 2 shows a schematic of a basic open flow system that uses compressed air as the heat transfer fluid. In the figure, compressed air is supplied from cylinders or from tanks that can be maintained at a defined pressure using air compressors and pressurestats. Gas boosters may also be used to reach higher pressures. Regulators are used to establish the system pressure, and back-pressure regulators are used to adjust the downstream pressure in order to "dial-in" the differential pressure across the heat exchanger, which can vary between zero (no-flow condition) and the difference between the regulated pressure and atmospheric pressure (full-flow condition). In-line heaters are used to increase the temperature of the air from near room temperature to high-temperature, and coolers are used to cool the gas exiting the heat exchanger down to a low enough temperature that it does not damage the valves and back-pressure regulators downstream. Sensors are installed that can measure the temperature, pressure, and flow-rate of the air before and after the heat exchanger, so that heat exchanger performance, pressure drop, friction factors, and other relevant information can be calculated. For a laboratory-scale system, the heaters may be electrically driven, and examples of in-line heaters capable of heating air to 700°C at 150 psig are shown in Figure 3. If steam is used instead of air, a lab-scale steam boiler or laboratory steam source can be substituted for one of the high-pressure air tanks. Multiple heat exchangers and other components could be tested with such a test bed as long as the gas input and output manifolds for the test article are manufactured to conform to the standard sizes and configurations established for the test bed. A co-current arrangement of flows is shown, but the inlet and outlet piping should be arranged to allow for co-current, cross-current, and counter-current flow.

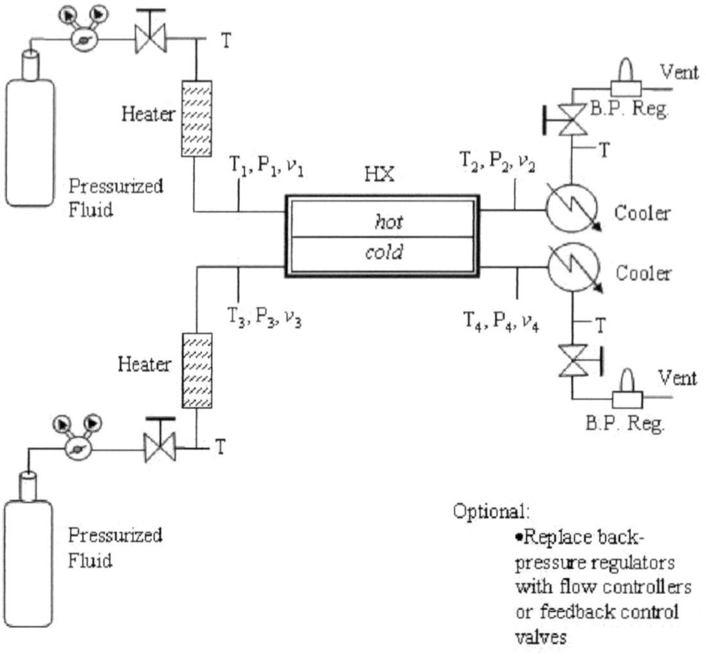

Figure 2: Once-through heat exchanger test apparatus.

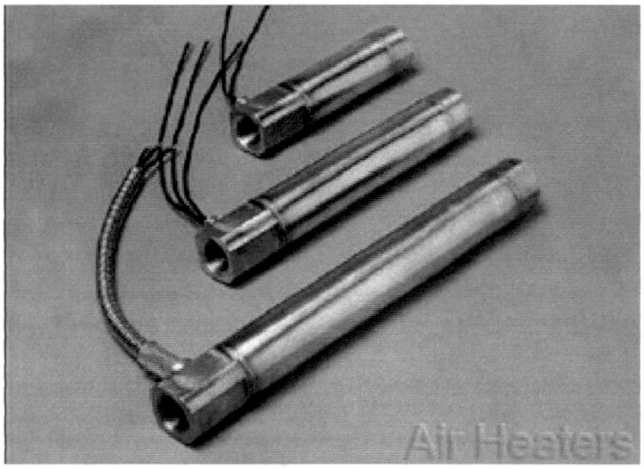

Figure 3: In-line air heaters operate up to 700°C and 120 psig (Heat Torch™200 by Farnum Custom Products, Inc.).

Partially Closed Loop

At a later stage of testing, one of the once-through air paths can be replaced with a closed helium loop to more closely approach the test conditions of interest. A schematic of a partially closed heat exchanger test bed is shown in Figure 4.

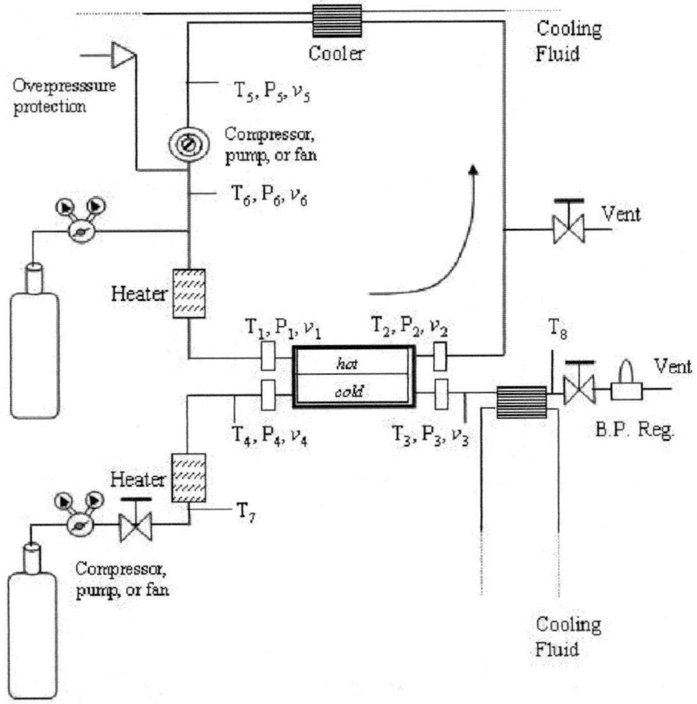

Figure 4: Partially closed heat exchanger test apparatus.

In Figure 4, the close loop is connected to a pressurized helium source, and a compressor is used to move the helium through the loop. Because there are few (inexpensive) compressors that can operate at temperatures as high as 800-900°C, a cooler and heater may be installed before and after the compressor to reduce the operating temperature to the compressor. Adding a cooler and heater to the loop reduces the thermal efficiency of the loop, but thermal efficiency is not the point of the test bed, and its purpose is to create fluid flow conditions that are suitable for a particular heat exchanger or component test. In addition to the compressor, temperature, pressure, and flow rate instrumentation is provided in all sections of the loop in order to monitor fluid flow conditions. A co-current arrangement of flows is shown, but the inlet and outlet piping should be arranged to allow for co-current, cross-current, and counter-current flow.

Two Closed Loops

The next level of sophistication of the heat exchanger test apparatus occurs when both once-through loops are replaced with closed helium loops. Such a configuration is shown in Figure 5. The second closed loop is a replicant of the first closed loop, and the two loops are operated independently. Helium is supplied batch-wise to the loops, and is only released from the loops when it is necessary to reduce system pressures for a test, or to remove or install a test article or other device in the apparatus. A co-current arrangement of flows is shown, but the inlet and outlet piping should be arranged to allow for co-current, cross-current, and counter-current flow.

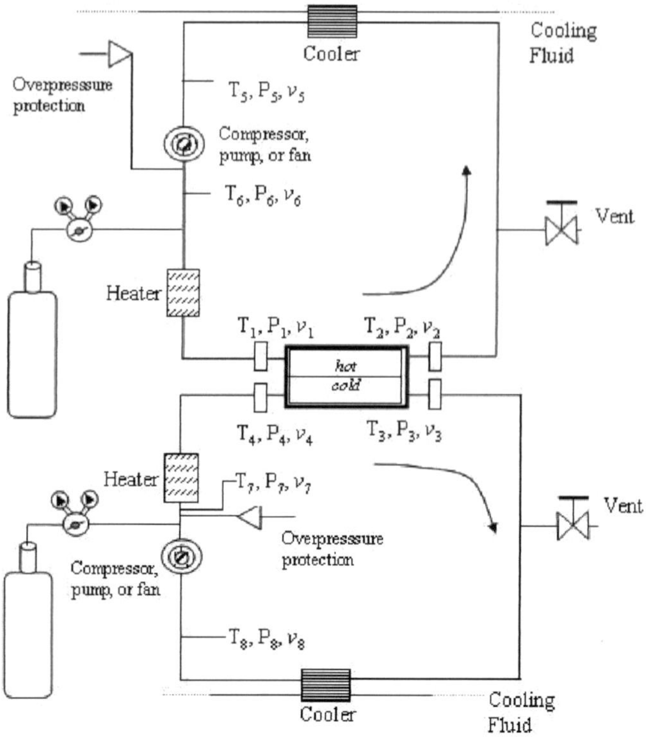

Figure 5: Fully closed heat exchanger test apparatus.

Similitude

The fluid and heat transfer properties of air and steam are different than high pressure helium at a given temperature and pressure, but the conditions of the experiments with air or steam can be changed to correspond to fluid conditions that are similar to or at least approaching what would have been realized had hot high-pressure helium been used instead. Though the individual properties of fluids will differ (density, viscosity, thermal conductivity, etc.), different fluids will behave in similar ways when certain dimensionless combinations of their fluid properties and the system geometries are similar. This similarity is called similitude [9]. In fluid and heat transfer systems, similarity is indicated by matching dimensionless numbers - Reynolds number (Re), Prandtl number (Pr), Graetz number (Gz), Stanton number (St), Nusselt number (Nu), and others. Testing of devices under similar conditions has been discussed by Kline [10], Bardet and Peterson [11], and Ingersoll et al. [12]. The Re, Pr, and Nu are related to each other through the Gz and St numbers (see Equations 1 and 2).

$$\mathrm{Gz} = \frac{\pi}{4}\mathrm{RePr}\frac{D}{L} \tag{1}$$

$$\mathrm{St} = \frac{\mathrm{Nu}}{\mathrm{RePr}} \tag{2}$$

In Equation 1, the parameters D and L are the diameter and length of the flow channel. Depending upon whether the flow conditions are laminar or turbulent, there are a number of correlations that use the Re, Pr, Gz, or St to calculate the Nu and, as a result, the heat transfer film coefficient. Texts on heat transfer should be consulted in order to determine which correlations are applicable to the test conditions of interest (e.g., laminar flow, turbulent flow, tubular versus planer geometries, etc.).

Next Steps

A high-temperature leak testing apparatus has been described, and concepts for once-through and close loop heat exchanger test beds have been offered that may be constructed at reduced capital cost in comparison to building a fully closed, pressurized, heated test facility for laboratory-scale heat exchangers and other components for the NGNP Project. Certainly, closed loops will be needed to advance the technological readiness of prototype heat exchangers, but the initial stages when technical risk is high, it is beneficial to have access to a relatively inexpensive test bed that can be used to screen heat exchanger concepts and to perform initial model validation under test conditions that are more easy to reach with simple equipment. Later, as hands-on experience grows and heat exchanger designs mature, the investment in closed loops can be made with less risk of failure due to component failure, poor heat exchanger design, and operator inexperience. It is recommended that such test systems be considered as stepping stones to larger and more complicated facilities that would, in time, be capable of providing test conditions approaching those of the projected nuclear plant design.

References

[1] (November 2007), *Next Generation Nuclear Plant: Preliminary Design Report*, INL/EXT-07-12967.

[2] Herring, S.J., O'Brien, J.E., Stoots, C.M., Hawkes, G.L., Hartvigsen, J.J., and Shahnam, M. (April 2007), "Progress in High Temperature Electrolysis Using Planar SOFC Technology," *Int. J. of Hydrogen Energy*, Vol 32, Issue 4, pp 440-450.

[3] Summers, W.A., Gorensek, M.B., and Buckner, M.R., (October 9-13, 2005), "The Hybrid Sulfur Cycle for Nuclear Hydrogen Production," *Proceedings of Global 2005*, Paper 097, Tsukuba, Japan.

[4] Schultz, K.R., Brown, L.C., Besenbruch, G.E., and Hamilton, C.J. (April 6-9, 2003), "Large Scale Production of Hydrogen Using Nuclear Energy for the Hydrogen Economy," GA-A24265. *Proceedings of the National Hydrogen Association 14th Annual Conference.*

[5] (2007), American Society of Mechanical Engineers (ASME), Boiler and Pressure Vessel Code, Sections III and VIII.

[6] Extrand, C.W., Ware, D., Kirchgessner, J., and Schlutter, J. (2008), "Quantifying Leak Rates in Gas Handling Systems," *Ind. Eng. Chem. Res.*, Vol 47, pp 1304-1309.

[7] Sherman, S.R., and Chen, Y. (2008) "HX Testing Requirements and Facility Needs for the NHI/NGNP Project," WSRC-STI-2008-00152, Savannah River National Laboratory.

[8] See USGS web site at http://minerals/usgs.gov/minerals/pubs/commodity/helium/mcs-2008-helium.pdf.

[9] Hughes, W.F. and Brighton, J.A. (1999), *Schaum's Outline of Theory and Problems of Fluid Dynamics*, McGraw-Hill Professional.

[10] Kline, S.J. (1988), *Similitude and Approximation Theory*, Springer-Verlag.

[11] Bardet, P. and Peterson, P.F. (October 2-6, 2005), "Design of Scaled Integral Experiments for High Temperature Liquid Salt and Helium Fluid Mechanics and heat Transfer," *Proceedings of the Eleventh International Topical Meeting on Nuclear Reactor Thermal Hydraulics*, NURETH 11, Avignon, France.

[12] Ingersoll, D.T., Forsberg, C.W., MacDonald, P.E., et al. (February 2007), *Trade Studies for the Liquid-Salt-Cooled Very High-Temperature Reactor: Fiscal Year 2006 Report*, ORNL/TM-2006/140.

OVERVIEW OF THE U.S. DOE HYDROGEN FROM COAL PROGRAM

C. Lowell Miller, Director, Office of Sequestration, Hydrogen, and Clean Coal Fuels
Mark Ackiewicz, Hydrogen and Clean Coal Fuels Program Manager
United States Department of Energy, Office of Fossil Energy

1. Introduction

Two of the major concerns confronting the energy sector today and for the future are energy security and the environmental impact of energy use, including the potential for reduction and stabilization of greenhouse gas emissions and atmospheric concentrations. As shown in Figure 1, imports of liquid fuels will continue to remain a large part of the United States' total liquid fuels supply [1,2] even as actions to increase the amount of renewable fuels and vehicle fuel economy as a result of the Energy Independence and Security Act of 2007 are implemented.

Figure 1: U.S. Liquid Fuels Supply (1950-2030)

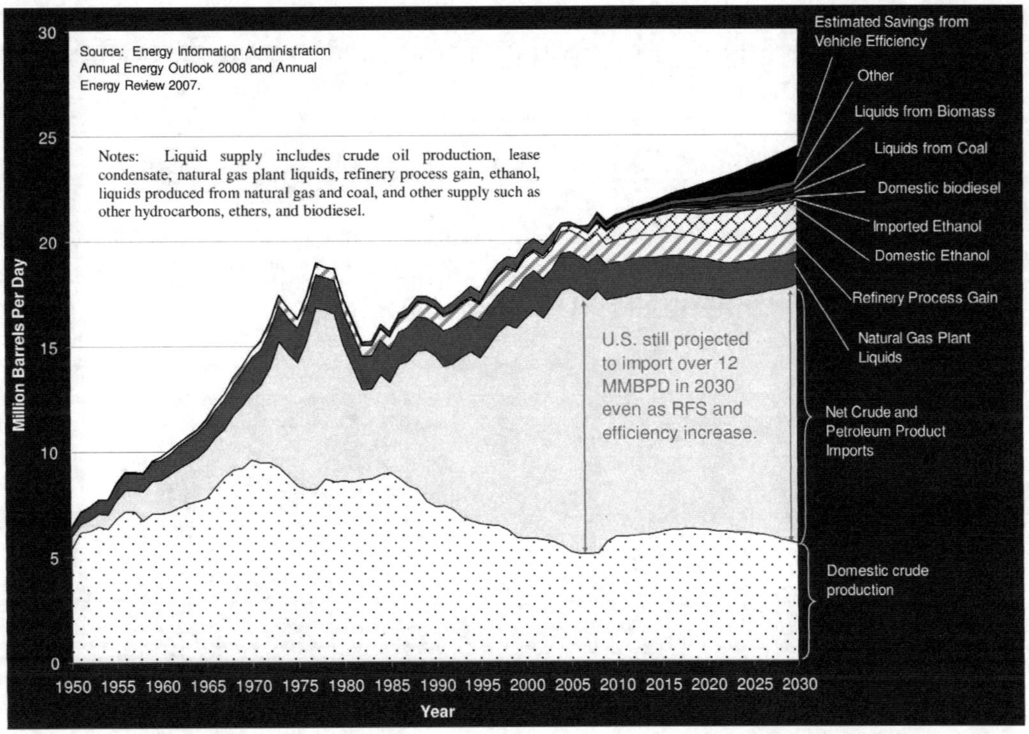

Hydrogen has the potential to play a significant role in America's energy future because it can be produced from a variety of domestic feedstocks and its use produces little or no emissions. Realizing this potential, the U.S. government launched the Hydrogen Fuel Initiative in 2003.

In support of this initiative, the Hydrogen from Coal Program was launched in fiscal year 2004 (FY2004) to perform the necessary research, development, and demonstration (RD&D) activities to develop technology to cleanly produce, deliver, store, and utilize hydrogen produced from our abundant domestic coal resources. In addition to managing its respective

RD&D portfolio, the Program also cooperates on joint efforts with other DOE and FE offices to coordinate and integrate planning efforts on key activities such as the Hydrogen Posture Plan.

2. Benefits of Hydrogen Production from Coal

As a pre-eminent primary source of energy, coal is an abundant domestic resource, with the United States possessing hundreds of years of supply at current demand levels. The production of hydrogen from coal for use in fuel cell vehicles (FCVs) in the transportation sector has the potential to reduce U.S. reliance on foreign imports of petroleum. It is estimated that, if used hydrogen from coal was used in 100 million FCVs it could reduce petroleum demand in the transportation sector by three million barrels per day.

An analysis of the NOx, SOx, CO_2 emissions from hydrogen from coal plants was performed that evaluated and compared the environmental benefits of hydrogen produced from coal when used in FCVs with gasoline-powered internal combustion engine vehicles (ICEVs) and hybrid electric vehicles (HEVs) [3]. The analysis was done on a systems basis that calculated the energy and emissions for resource extraction and transportation, conversion of the resource to fuel and fuel delivery (i.e., refining of petroleum to gasoline or processing of coal to produce hydrogen), and finally use in a vehicle. Figure 2 shows that, with carbon sequestration in the production of hydrogen from coal, net system emissions of CO_2 will be nearly eliminated compared to the internal combustion engine vehicle and hybrid electric vehicle systems with carbon sequestration. The analysis results also show that emissions of nitrogen oxides (NOx) and sulfur oxides (SOx) will be reduced.

Figure 2: Well-to-Wheels Emissions of Hydrogen from Coal and Use in Fuel Cell Vehicles Compared to Gasoline Use in Hybrid Electric and Internal Combustion Engine Vehicles (100 million vehicles)

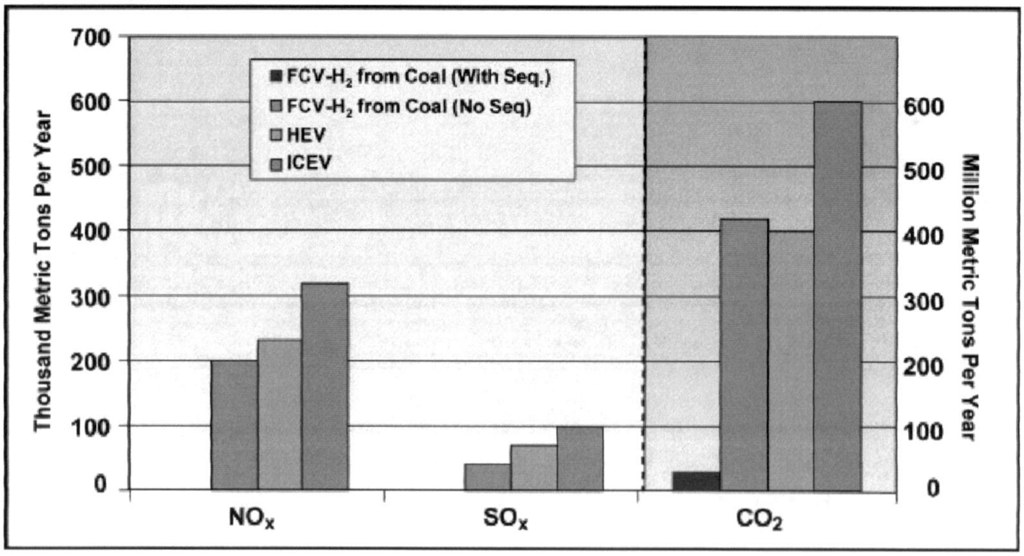

3. Hydrogen from Coal Production Process

A simplified flow diagram of the hydrogen from coal process is shown in Figure 3. The process begins by first gasifying coal to produce synthesis gas, a mixture of hydrogen, carbon monoxide, carbon dioxide and other gaseous compounds. The synthesis gas is cleaned of impurities and undergoes the water-gas shift reaction which converts water (steam) and carbon monoxide to hydrogen and carbon dioxide. Purified hydrogen is separated from this stream and the remaining tail gas can be combusted to produce power and the resulting CO_2 stream subsequently sequestered.

At the present time, no coal-fed facilities based on modern entrained gasification have been constructed that co-produce both hydrogen and electric power. Conceptual commercial plants have been simulated using computer models to estimate the technical performance and economics of a co-production plant based on current technology. Computer simulations have also been developed for conceptual plants that produce hydrogen and some excess power, based on advanced technologies that are presently not available for commercial deployment. The status of these advanced technologies varies. Some technologies are already close to commercialization and others are further back in the RD&D pipeline. Table 1 summarizes the information developed from three of these computer simulations discussed in the Mitretek Systems (now Noblis) Technical Report, *Hydrogen from Coal – July 2002* [4]. This study is currently being updated to reflect recent trends in materials of construction prices and engineering, procurement, and construction availability. Additionally, the updated study will evaluate the impact of technologies under development by the program.

Figure 3. Schematic of the Hydrogen from Coal Production Process

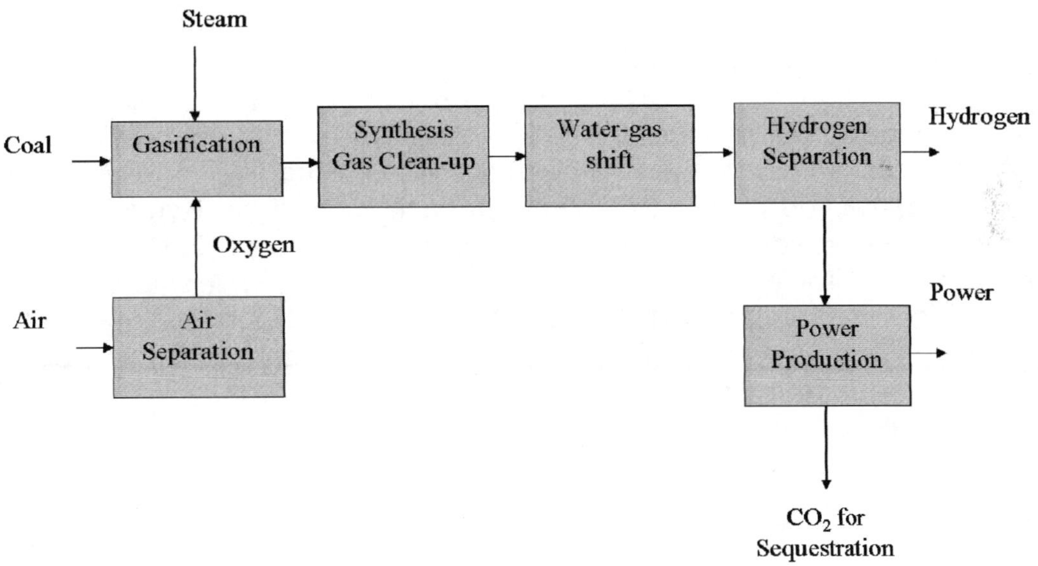

Table 1: Summary of Hydrogen from Coal Cases

	CASE 1	CASE 2	CASE 3
Carbon Sequestration	YES (87%)	Yes (100%)	Yes (100%)
Hydrogen (MMscfd)	119	158	153
Coal (Tons/day) (AR)	3000	3000	6000
Efficiency (%HHV)	59	75.5	59
Excess Power (MW)	26.9	25	417
Power Value (mils/kWh)	53.6	53.6	53.6
Capital ($million)	417	425	950
RSP of Hydrogen ($2002)	$/MMBtu ($/kg)	8.18 (1.10)	5.89 (0.80)

Notes: 1) Coal cost is $29/ton (and is assumed to de-escalate at 1.5 percent below general inflation), and the assumed plant capacity factor is 85 percent. 2) For carbon sequestration, the co-produced power is assumed to have a value of $53.6/MWh (megawatt-hour), based on an additional cost of power production from Natural Gas Combined-Cycle (NGCC) plants with sequestration of 18 mils/kWh (reference EPRI report 1000316). 3) For sequestration, it is assumed that $10 per ton of carbon is added for sequestration after the concentrated CO2 stream has been isolated, and the CO2 stream is compressed to 200 bars (2,900 pounds per square inch (psi)). 4) Financial assumptions used for these simulations: 25-year plant life; 67/33 percent debt/equity financing; 15 percent return on equity; 8 percent interest for a 16-year term; 3 percent inflation with coal de-escalation of 1.5 percent per annum below general inflation; 16-year double declining balance depreciation; 40 percent combined federal and state tax rate; 3-year construction with 50 percent output in start-up year; carbon sequestration cost of $10/ton.

4. Hydrogen from Coal Program

The challenges to a sustainable supply of energy and the benefits of hydrogen as an alternate fuel serve as drivers for the Hydrogen from Coal Program. The Hydrogen from Coal Program has an established history, based on prior activities, of performing research on novel technologies to produce hydrogen from coal. The Program has RD&D activities focused on the production and delivery of hydrogen in large, central, single product and coproduct facilities, production of hydrogen-rich liquid fuels, and production of substitute natural gas (SNG) from coal. Central production of hydrogen relies upon gasification of the coal with steam and oxygen to produce synthesis gas, a mixture of hydrogen, carbon monoxide, and trace amounts of other components. The synthesis gas can be further processed to remove impurities and produce additional hydrogen. As noted, central production also offers the potential of coproduction of hydrogen and electric power with capture and sequestration of CO_2.

An alternative production pathway investigates the production of hydrogen-rich liquid fuels and SNG that have the advantage of using the existing petroleum product and natural gas delivery and storage infrastructures with little to no modification. Liquid fuels and SNG can be delivered either to individual filling stations where they can be reformed on-site, or to sub-central hydrogen production locations to produce pure hydrogen.

The program is also investigating the use novel materials, such as metal-organic frameworks, as a potential option for storing hydrogen on-board vehicles, and novel engine systems that may potentially utilize hydrogen or hydrogen-natural gas mixtures in vehicular or stationary applications. Activities in hydrogen storage and utilization will be completed by the Program in FY2009.

More details about the Program can be found in FE's *Hydrogen from Coal Research, Development, and Demonstration Plan, External Draft for Review – September 2008* [3].

4.1 Budget

Past budget appropriations and the current budget request for the Hydrogen from Coal Program are shown in Table 2. The Program initiated activities under the Hydrogen Fuel Initiative in FY2004 and funds are typically leveraged in cost-sharing activities with project partners, allowing greater project scope and participation.

Table 2: Past and Recent Budget Data for the Hydrogen from Coal Program

Program	FY04	FY05	FY06	FY07	FY08*	FY09**
Hydrogen from Coal	5.0	17.1	28.7	22.1	24.8	10.0

* Includes funding for hydrogen activities and coal-biomass research as directed by Congress.
** FY09 budget request.

4.2 Central Station Production of Hydrogen

<u>*Advanced water-gas shift reaction systems*</u>

Partial oxidation (i.e., gasification) of coal and other carbon-based solid/liquid feedstocks produces a synthesis gas with a composition ranging from 30 to 45 percent H_2, 35 to 55 percent carbon monoxide (CO), 5 to 20 percent CO_2 (dry basis), and trace amounts of other components. If the H_2 to CO ratio of the syngas from the gasifier is not appropriate for the synthesis of fuels or chemicals, the ratio can be adjusted using the WGS reaction. This reaction increases the concentration of hydrogen in the synthesis gas by converting CO and H_2O to CO_2 and H_2. When coupled with an appropriate separation technology, high yields of high-purity hydrogen are produced.

The WGS reaction is reversible, with the forward WGS reaction being mildly exothermic. Conversion to H_2 and CO_2 is thermodynamically limited and favored at lower temperatures. Higher temperatures improve the rate of reaction, but lower the yield of hydrogen. In order to achieve high yields at high rates of reaction, the reaction is typically carried out in multiple adiabatic reactor stages, with lower reactor inlet temperatures in the latter stages. The yield also may be improved by using excess steam or by removing hydrogen from the reactor to shift the WGS equilibrium to produce more hydrogen. Steam also is used to minimize undesirable side reactions that compete with the WGS reaction.

Hydrogen separation

Current hydrogen recovery methods typically employ pressure swing adsorption (PSA), cryogenics, or membrane systems. Each of these technologies has limitations: PSA typically recovers less of the feedstream hydrogen and is limited to modest temperatures; cryogenics is generally used only in large-scale facilities with liquid hydrocarbon recovery, because of its high capital cost; and current membrane systems are susceptible to chemical damage from H_2S and aromatics as well as having limited temperature tolerance.

There are several gas separation technologies under development by the Program that could separate constituents of the synthesis gas (syngas). These technologies are advanced membrane separation systems, advanced CO_2 separation systems, polishing filters, adsorption/solvent systems, and advanced concepts such as process intensification or chemical looping. The Hydrogen from Coal Program seeks to develop technologies to improve the separation of H_2 and/or CO_2 from synthesis gas streams that will reduce capital and operating costs and improve thermal efficiency and environmental performance.

Advanced membrane separation systems

Reductions in cost, improved efficiency, and simplified systems are potentially possible with advancements in hydrogen membrane separation technologies. Membranes can be classified as organic, inorganic, or hybrid (a mixture of organic and inorganic materials). Within each of these classes, membranes can also be characterized based on their properties. Although there are many types of membranes, the Hydrogen from Coal Program has focused its efforts on microporous membranes, pure metallic membranes, and hydrogen-permeable ceramic-metallic (i.e., cermet) membranes. Desirable characteristics of separation membranes are high hydrogen flux at low pressure drops; tolerance to contaminants, especially sulfur; low cost; and operation at system temperatures of 250 to 500°C. The pressure drop across the membrane is a critical aspect of membrane technologies. Because hydrogen produced at central plants will ultimately be needed at high pressures to both deliver and store adequate volumes, it is critical that this pressure drop be minimized, otherwise expensive re-compression may be needed which could negate any competitive advantage of membranes compared to other technologies.

Advanced CO_2 separations systems

Another method to separate H_2 mixed with CO_2 is to remove the CO_2 from the process stream. CO_2 currently is separated from syngas through commercially available CO_2 absorption systems. These are continuous scrubbing systems that typically are available as three basic types: chemical, physical, and hybrid. All the processes operate in essentially the same manner by scrubbing the mixed gas in absorption towers to collect the CO_2, and then regenerating the solvent and releasing the CO_2. After separation, the CO_2 stream is dried and compressed to approximately 1,600 pounds per square inch (psi), which results in liquid CO_2 that is transported to a utilization site (e.g., enhanced oil recovery) or to a future sequestration site.

Advanced CO_2 separation technologies have as their objective the efficient, low-cost removal of CO_2 and other trace impurities from the H_2-CO_2 mixtures. Reverse selective membranes, such as nanostructured, polymeric membranes, show potential to achieve the desired selectivity for CO_2 separation. CO_2 sorbents and hydrates are other options that selectively remove CO_2 from mixed gas streams. Sorbents work by adsorbing CO_2 gas molecules onto the surface of a solid. CO_2 hydrates remove CO_2 by forming a crystalline lattice of water molecules around the gas molecule, subsequently trapping the molecule.

Polishing filters (ultra-clean hydrogen purification systems)

Hydrogen produced from coal can be used for various applications, including transportation, gas turbine and solid oxide fuel cell (SOFC) power generation, and crude oil refining. The hydrogen purity requirement is dependent on the applications in which it is used. Proton exchange membrane (PEM) fuel cells, which are the preferred technology for use in FCVs, currently require the stringent standard of less than 1 ppm of CO, and very low levels of sulfur that may be limited to 10 to 50 parts per billion (ppb) levels or lower. A polishing filter device using an absorbent may be required to meet the particular application requirement. For coal-based hydrogen production, RD&D efforts have to identify and define the preferred absorption material or materials and conditions to successfully remove the trace impurities to meet the standards, with more focus on the sulfur removal process.

Advanced adsorption/solvent systems

Advanced adsorption (i.e., advanced PSA) and other solvent systems have the potential to improve current hydrogen separation technologies. These technologies can help lower the cost of current hydrogen separation from large centralized coal plants until membrane technologies become commercially available. These technologies may include novel catalysts, adsorbers, or solvents that make current technologies more efficient, improve environmental performance, increase operating capacity, reduce operating and maintenance costs, or improve processing.

Advanced concepts

The Hydrogen from Coal Program is investigating advanced concepts for producing and separating hydrogen such as process intensification. Process intensification is the concept of developing novel technologies which, when compared to current technology, identifies dramatic improvements that lead to more compact, energy efficient, and lower cost technologies. As related to hydrogen production from coal, these concepts could be a "one-box" process that combines synthesis gas cleanup, the WGS reaction, and hydrogen separation. Other advanced concepts include new process control methods or novel concepts that integrate alternative energy sources into the hydrogen from coal production facility. These advanced concepts will require long-term research efforts before they are ready for larger-scale development, but could significantly improve the production of hydrogen from coal. One of the more developed advanced concepts being investigated is the integration of the WGS reaction with hydrogen separation. Chemical looping is another advanced concept for producing and separating streams of hydrogen and CO_2 from mixed gases and trace constituents.

4.3 Alternate Hydrogen Production Pathway

The Hydrogen from Coal Program is also pursuing an alternative pathway to produce hydrogen from coal by producing liquid fuels or SNG as an intermediate step. The Office of Fossil Energy has been a leader in researching, developing, and demonstrating the production of synthesis gas and subsequent conversion to liquid fuels and chemicals. FE's research and development program has included the $213 million Liquid Phase Methanol (LPMEOH) demonstration project, led by Air Products and Chemicals, Inc. (APCI), which demonstrated on a commercial scale the production of methanol and dimethyl ether from coal-derived synthesis gas.

Liquids and SNG are characterized by high weight content of hydrogen and can be transported in either the existing liquid fuel or natural gas delivery system. They can subsequently be reformed closer to the end-user in sub-central or distributed locations. Some of the processing steps used to produce hydrogen in the central production pathway are similar for the alternative pathway. For example, coal needs to be gasified to produce synthesis gas for both pathways. However, for liquid fuels and SNG, the synthesis gas is then converted into liquid fuels via the synthesis gas to methanol or the Fischer-Tropsch (F-T) synthesis process or into SNG through methanation. Because these plants are centrally located, they can capture any of the CO_2 that is produced during these synthesis processes. The liquid fuels and SNG can be delivered to sub-central or distributed sites by the existing liquid fuel or natural gas delivery systems. Once at the sub-central or distributed plant, technologies such as steam methane reforming and autothermal reforming can produce hydrogen from the liquid fuel or SNG. There is potential to use advanced technology for the reforming of liquid fuels and SNG to produce hydrogen, which would take advantage of process intensification with a corresponding reduction in product cost. For sub-central plants, final delivery of the hydrogen to individual filling stations can be made using hydrogen tube trailers.

Two key benefits of this pathway are use of domestic coal as a hydrogen feedstock and use of existing delivery infrastructure, thus eliminating the need for construction of a new, capital-intensive hydrogen infrastructure. However, the trade-offs between capital investment in infrastructure and the addition of further processing needs to be evaluated. The cost, efficiency, and benefits associated with these pathways have to be evaluated on a system basis for comparison to other possible hydrogen system pathways.

Systems analysis studies of advanced conceptual plants that co-produce liquid fuels and electric power have projected that liquid fuels can be produced in these advanced facilities for about $40 to $50 per barrel. Site-specific studies have also shown that SNG can be produced from coal for about $6/MMBtu to $8/MMBtu based on various plant designs and configurations. Although significant improvements in the cost of liquid fuels and SNG production have been made in the past, additional improvements in production and reforming can enable this hydrogen production pathway to become an economic alternative.

5. Accomplishments and Progress

The Hydrogen from Coal Program has successfully transitioned from its initial start-up in FY2004 to full operations. The Program has been actively soliciting proposals from industry, universities, national laboratories and other organizations to help the program achieve its goals in support of the HFI. As of September 2008, the program has 34 projects that are conducting research in a wide number of areas (Table 3).

Table 3: Hydrogen from Coal Research Projects

Research Area*	Number of Projects
Membrane research	10
Membrane reactors & process intensification	5
CO_2 removal	2
Polygeneration	3
Liquid H_2 carriers	4
SNG production and reforming	2
Storage	3
Utilization	5
TOTAL	**34**

**Complementary projects are supported by the Gasification and Sequestration Programs.*

The Hydrogen from Coal Program is making progress towards achieving some of its goals, milestones, and technical targets. For example, in the area of hydrogen separations, early laboratory-scale research has demonstrated significant progress towards the Program's long-range targets. Additionally, some of these projects are beginning to advance to larger scales of development to obtain engineering data. Table 4 shows the progress towards the hydrogen separation technical targets by the Program's research activities.

NETL's Office of Research and Development (ORD) has also established a high-pressure hydrogen test facility that is currently testing membranes developed by the Program's research contractors' to verify their results. The facility has performed molecular chemistry and laboratory testing of sulfur-resistant palladium alloys for water-gas shift (WGS) membrane reactors. Testing has shown:

- that sulfur resistance has been observed for palladium-copper alloys in the presence of H_2S;
- corrosion and/or catalytic poisoning play a significant role in the permeability decreases observed in the presence of H_2S;
- H_2 permeability of palladium-copper alloys is strongly dependent on crystalline structure; and
- Enhanced CO conversion has been demonstrated in catalyst-free palladium-based membrane reactors at elevated temperatures in simulated coal-derived synthesis gas.

Table 4: Progress Towards Separation Technical Targets

Performance Criteria	Current Status (a)	2010 Target	2015 Target
Flux (b)	>300	200	300
Temperature, ºC	300–400	300–600	250–500
S Tolerance	Yes (preliminary)	Yes	Yes
Cost, $/ft^2	<200	100	<100
WGS Activity	Yes	Yes	Yes
ΔP Operating Capability (c)	1,000	Up to 400	Up to 800 to 1,000
Carbon Monoxide Tolerance	Yes	Yes	Yes
Hydrogen Purity (d)	>99.999%	99.5%	99.99%
Stability/Durability (years)	<1	3	5

[a] Current status is based on best available membrane technology to date (2006) that meets the targets under laboratory conditions.

[b] ft^3/hour/ft^2/100 psi ΔP (hydrogen partial pressure basis)

[c] ΔP = total pressure differential across the membrane reactor (psi)

[d] Polishing filters may be needed downstream of the separation system in order to remove final traces of CO, sulfur, and other impurities to meet PEM fuel cell requirements. These targets exclude the effect of polishing filters.

** Technical targets are for membrane types described previously; research that is currently supported by FE and NETL Research on other membranes is not precluded if the potential flux, cost, and tolerance to impurities are promising. Research is also encouraged on advanced solvent and adsorption technologies to separate hydrogen. Reverse selective systems that separate CO2 also are promising. The technical targets for hydrogen membranes relate to hydrogen from coal technology in which delta P will be around 100 psi and the membrane will require resistance to contaminants (CO and H2S). Technical targets for hydrogen membranes that are included in the EERE HFCIT RD&D Plan are for systems that operate at lower delta P and have less contaminants.*

6. Future Efforts

Future efforts will emphasize continued production and purification of hydrogen from coal, particularly membrane separations. These activities will focus on scale-up of technologies to further advance and validate laboratory results. Additionally, the program will continue to re-evaluate its program goals in consideration of budget projections and modifications to key activities.

7. References

1. Energy Information Administration (EIA). Annual Energy Outlook 2008. http://www.eia.doe.gov/oiaf/aeo/index.html
2. EIA. Annual Energy Review 2007. July 2008. http://www.eia.doe.gov/emeu/aer/
3. DOE Office of Fossil Energy, Office of Sequestration, Hydrogen, and Clean Coal Fuels. Hydrogen from Coal RD&D Plan – External Draft for Review. September 2008.
4. Mitretek Systems. Hydrogen from Coal, Mitretek Technical Paper MTR 2002-31. July 2002. http://www.netl.doe.gov/technologies/hydrogen_clean_fuels/refshelf/pubs/Mitretek%20Report.pdf

635b Hydrogen Production from Nuclear Energy; The DOE-NE Perspective

Stephen Kung and Carl J. Sink Jr. Nuclear Hydrogen Initiative, U.S. Department of Energy, 19901 Germantown Road, Germantown, MD 20874

This presentation will describe the U.S. Department of Energy, Office of Nuclear Energy vision for the role of nuclear energy in a future Hydrogen Economy in the United States.

635c Large Scale Hydrogen Production Using Solar Energy -- Status, Challenges, and Plans

Richard Farmer, Office of Energy Efficiency and Renewable Energy, U.S. Department of Energy, Hydrogen, Fuel Cells, and Infrastructure Technologies Program, 1000 Independence Avenue, SW, Washington, DC 20585

To improve U.S. energy security and to supplement biofuels as a replacement for petroleum in light duty vehicles, a low cost, reliable, and renewable source of energy will be required. One option that is being investigated by the U.S. Department of Energy (DOE) is the production and use of hydrogen in hydrogen fuel cell vehicles. Projections show that near term production of hydrogen at the fueling station using steam-methane reforming can meet the DOE cost targets of $2.00 - $3.00/gge, untaxed and delivered. Longer term centralized production of hydrogen from coal with sequestration and biomass gasification along with nuclear energy provide additional opportunities and are under investigation. Alternate approaches to these centralized hydrogen production technologies include solar energy for water splitting using photobiological, photoelectrochemical, and thermochemical processes. This presentation will discuss the current status of each of these solar technologies, the challenges that each must overcome, and the approach that is being taken to address the challenges and to advance the technologies closer to commercialization.

635d Biomass Gasification for Hydrogen Production

Richard L. Bain, National Bioenergy Center, National Renewable Energy Laboratory, 1617 Cole Blvd., MS 3322, Golden, CO 80401

Analysis of hydrogen production using indirect gasification for syngas production followed by gas cleanup, shift reaction, and hydrogen separation has the potential to meet DOE plant gate cost targets of $1.60/gallon gasoline equivalent in 2012 and $1.10/gallon gasoline equivalent in 2017. These analyses are base on pilot gasification data developed in the 1980's.

In this presentation the impact of new correlations based on recent pilot scale gasification testing will be discussed, including the influence of gasification temperature, steam to biomass ratio, and gasifier residence time. The presentation will include a discussion of updated ASPEN simulation and economic analyses comparing the original model and the update model. The presentation will also compare differences in technology assumptions and costs for hydrogen from coal and biomass.

694a A Preliminary Physical-Chemical Analysis of Steam Reforming of Biogas from a Small Wastewater Treatment System of a University Campus

Antonio Carlos Caetano Souza and José Luz Silveira. Energy, UNESP (São Paulo State University), Av Ariberto Pereira da Cunha, 333, Guaratinguetá, 12514-410, Brazil

The need to diminish environmental impact caused by energy generation and consumption has leaded researches concerning renewable resources such as biogas.

A small and low-cost wastewater treatment system with biogas production was developed in a campus of São Paulo State University, campus of Guaratinguetá. This system is able to receive and to treat part of the wastewater produced in this campus, diminishing costs. Previously, the overall volume of this residue has piped by a state-owned company.

Beyond treating wastewater, an attention concerning the water produced during this treatment (reclaimed water) and the produced biogas must taking into account. This type of water (which contains nutrients for the plants) could be used in the gardens. The biogas could be used for energy purposes in own campus.

The utilization of biogas as a feedstock for biohydrogen production has be widely researched due to its low environmental impact and due to its availability, since this fuel could be produced utilizing various feedstocks through various technologies and processes. In this case, due to the small amount of produced biogas and hence, small amount of hydrogen, its use as energy carrier is imperative.

Recently, a project was approved under support by FAPESP, a state-owned company responsible by financial support of projects performed in the Brazilian state of São Paulo. Firstly, a theoretical analysis was developed to evaluate the content of hydrogen to be produced through steam reforming. The choice of this technology of hydrogen production is due to the large experience devoted by Energy Systems Optimization Group, maintained in this campus. This group has developed recently technologies and process of hydrogen production utilizing inclusively other sources such as natural gas and ethanol. Another motive to choose this technology is the ease of project and development and its low cost of installation. Other measurements such as exergy efficiencies and irreversibilities have contributed to determine the steam reforming process if compared with other processes such as partial oxidation and autothermal reforming. However, due to the high amount of carbon dioxide (CO2) in the produced biogas, and due to the high cost necessary to extract it, a steam reforming process associated with dry reforming (where the own CO2 is utilized to react with methane gas (CH4), the principal constituent of biogas) is suggested. Both reforming processes are accepted and no more additional process is necessary since the volume of CO2 contained into the biogas is smaller if compared with CH4. The volume of constituents was 54% CH4 and 40% CO2 before H2S purification, and 61.8% CH4 and 34.4% CO2 after purification. Some traces of O2 and N2 were encountered in the biogas in both cases however these small amounts are negligible and no problems during hydrogen production are foreseen. The amount of encountered H_2S was 2.8% before purification and 1.2% after purification process. This process is performed through low-cost devices which utilize small pieces of iron extracted from machining processes.

In this work a physical-chemical, exergetic and economic were developed as mean to determine the more optimized conditions for hydrogen production. The steam reforming and dry reforming are endothermic processes. Due to this, high temperatures of hydrogen production are foreseen and hence a major amount of fuel as heat source is necessary. For a biogas production of 9 Nm3 per day (that is the capacity of production of biogas in the wastewater treatment system), the reforming process in the best conditions could produce about 1,3 kg H_2 per day, which could generate about 15-20 kWh electricity if a PEMFC (Proton Exchange Membrane Fuel Cell) is considered. A performance of 100% for hydrogen production process was adopted, and following some recent works, the electric efficiency of a PEMFC could attain 40-50%.

The suggested thermodynamic conditions detected in physical-chemical and exergetic analysis was 600° C and 1 atm, where lower irreversibilities, high exergetic and rational efficiencies and high conversion efficiency low could be encountered. The performance of hydrogen production increases as soon as temperature of reaction also increases, however, at temperatures grater than 600°C, the additional hydrogen to be produced is low if compared with high amount of energy necessary to the process. In contrary to temperature, as soon as pressure during hydrogen production increases, the performance of reaction decreases.

Connected with both analyses, an economic analysis was developed to evaluate its viability. Some findings were considered such the possibility to utilize electricity and the own biogas as heat source to perform the reactions, the annual period of utilization of developed hydrogen production system of 5000 to 7000 hours, maximum interest rate of 20% per annum (considering the Brazilian conditions), and maximum payback of 20 years. The lowest cost detected of hydrogen produced was 0.27126 and 0.22323 US$/kWh whether electricity and biogas as heat sources are utilized, respectively. In the last case, an additional cost associated with biogas was not considered. This cost should be added in some cases due to the utilization of equipments (such as H_2S purification system) to maintain biogas production only although this cost is low and do not become the process prohibitive.

The introduction of emission trading schemes such as carbon credits have become possible the introduction of environmentally-friendly technologies, in special for technologies which allow the capture of greenhouse gases such as biogas. The evaluation of the produced biogas volume and the amount of greenhouse gases emitted by hydrogen production process allow to determine the value of allowance, which could be utilized to the infra-structure, diminishing the costs of produced hydrogen and hence, becoming this technology more-competitive. With cost of allowances attaining US$ 40 per ton of CO_2-equivalent, the installation of a hydrogen production system with no-additional funds is possible.

USE OF TECHNOLOGY READINESS LEVELS AS A DEVELOPMENT TOOL

Steven R. Sherman, Savannah River National Laboratory, Aiken, South Carolina, USA

Introduction

Technology Readiness Levels (TRLs) are a means to categorize the relative technical maturity of components, sub-systems, or systems. Technological or technical maturity is a measure of performance, reliability, durability, and operating experience, and more technologically mature components (better performing, more reliable, more durable, greater operating experience) are rated higher than experimental or prototype components or systems. In this definition, an understanding of the supporting terms is important, and these also have specific meanings. Performance is the component's or system's ability to perform the task for which is intended. Reliability is a measure of repeatability or consistency, or the ability of a component or system to perform at a particular level for sustained periods of operation. Durability is the ability of the component or system to withstand stresses and wear without a degradation or failure in performance. Increased operating experience is correlated with higher technical maturity because it provides more data to support more accurate assessments of performance, reliability, and durability, which in turn support optimization of design and functionality.

Various government agencies and research programs are now using TRLs to assess the technological maturity of components and systems in order to drive the technology development process. The National Aeronautics and Space Administration (NASA), the U.S. Department of Defense (DoD), and the DOE Office of Environmental Management (DOE-EM) employ an integer rating system that spans from 1 to 9, with the lowest number corresponding to "basic principles observed" and the highest number corresponding to "mission proven." The Next Generation Nuclear Plant (NGNP) Project, which is funded by the DOE Office of Nuclear Energy (DOE-NE), is adopting the NASA and DoD systems to govern its work, while the DOE Nuclear Hydrogen Initiative (also DOE-NE) and the DOE Hydrogen Program (DOE Office of Energy Efficiency and Renewable Energy, EERE) are exploring systems that span 1 to 6 and 1 to 8, respectively.

By themselves, TRLs are abstractions that help facilitate the communication of complex ideas between developers and project stakeholders; that is, between scientists and engineers, who tend to be highly technical, and program managers and sponsors, who tend to be less technical. For example, instead of a developer discussing with a program sponsor whether the latest design of Device X is operating at 23% efficiency with a mean time between failures of 650 hours, a developer can instead reference a TRL in the discussion. Since a common understanding of what that TRL means in terms of performance, reliability, durability, and operating experience will have been previously established, less confusion would result. If there is only one component or system under development by the program, then the facilitation is moot, but it becomes much easier to use TRLs in such discussions rather than direct technical data when the number of components or systems under development is greater than just a few, and each component or system under development is designed for a different purpose.

Additional value is drawn from TRLs by virtue of the programmatic infrastructure needed to support the TRL assessment system. Assigning a TRL requires the establishment of an objective entry and assessment process, identification of expected component or system performance at each stage of development, and application of quality assurance processes for technical data and other supporting records. As a natural extension, such processes also almost automatically lead to the identification and generation of information that can be used to support intellectual property claims.

The use of TRLs is a risk reduction strategy. A component's or system's TRL is loosely correlated with the capital and operating cost required to perform a test in order to verify that a particular component or system is at a particular TLR. For example, it is usually cheaper to perform a test of a small prototype component in the laboratory in order to provide proof-of-principle than it is to perform long duration testing of a full-sized component in an operational or plant environment. TRLs are also indicative of risk, and lower TRLs, due to lack of data or inherent design problems, are correlated with a higher probability of failure. In a development process, only successful tests lead to higher technical maturity, while failed tests either lead to lost investment or higher investment if tests must be repeated. Therefore, greater programmatic efficiencies (lower overall costs) may be realized if the capital and operating investments made into testing a component or system are scaled appropriately to the current testing phase and target TRLs.

Successful use of TRLs by a development program requires the sustained support of program management, and active management of the supporting processes.

In this paper, existing TRL systems are discussed. The mechanics of managing and using a TRL system are described. Further details about how components, sub-systems, or systems may be initially assessed (the entry process) and promoted are provided. The importance of building development case files and maintaining records is highlighted. Also, a more thorough discussion of communications and risk is undertaken in order to highlight the advantages of using TRLs as a tool to help manage large technology development projects.

TRL Scales

The model system is one established by NASA in the 1990's, and is used to categorize the relative technological maturity of components, sub-systems, and systems [1, 2]. Definitions of the NASA TRLs are given in Table 1.

After NASA adopted its system, other parts of the U.S. government adapted the NASA TRL definitions to their own needs, and now DoD [3] and DOE-EM [4] use similar scales to the NASA scale. In these other systems, the number of TRLs are the same, and the higher-level definitions are the same (shown in bold in Table 1), but the individual interpretation of the higher-level definitions (shown in regular text in Table 1) are altered to meet the needs of the individual research and development programs.

Other government-sponsored research and development programs are following in the path prepared by NASA, DoD, and DOE-EM and are working to develop their own TRL systems, but are changing the number of levels employed to assess technological readiness. The DOE Nuclear Hydrogen Initiative (NHI) out of the DOE Office of Nuclear Energy (DOE-NE) has offered a system that has six levels [5]. The DOE Office of Energy Efficiency and Renewable Energy (DOE-EERE) Hydrogen Program is working on a system with 8 levels. The Next Generation Nuclear Plant (NGNP) Project, also out of the DOE-NE, is developing a system with 10 levels. These other scales are not a departure from the original NASA model, but instead combine or expand elements of the NASA scale to provide more or less emphasis on different stages of development. Mapping the definitions of the TRLs in these scales to each

Table 1: **NASA TRL System**

Level	Description
1	**Basic Principles Observed and Reported.** Lowest level of technology readiness. Scientific research begins to be translated into applied research and development. Examples might include paper studies on a technology's basic properties.
2	**Technology Concept and/or Application Formulated.** Invention begins. Once basic principles are observed, practical applications can be invented. Applications are speculative and there may be no proof or detailed analyses to support assumptions. Examples are limited to analytical studies.
3	**Analytical and Experimental Critical Function and/or Characteristic Proof of Concept.** Active research and development is initiated. This includes analytical studies and laboratory studies to physically validate analytical predictions of separate elements of the technology. Examples include components that are not yet integrated or representative.
4	**Component and/or Breadboard Validation in Laboratory Environment.** Basic technological components are integrated to establish that they will work together. This is relatively "low fidelity" compared to the eventual system. Examples include integration of *ad hoc* hardware in the laboratory.
5	**Component and/or Breadboard Validation in Relevant Environment.** Fidelity of breadboard technology increases significantly. The basic technological components are integrated with reasonably realistic supporting elements so it can be tested in a simulated environment. Examples include "high-fidelity" laboratory integration of components.
6	**System/Subsystem Model or Prototype Demonstration in a Relevant Environment.** Representative model or prototype system, which is well beyond that of TRL 5, is tested in a relevant environment. Represents a major step up in a technology's demonstrated readiness. Examples include testing a prototype in a high-fidelity laboratory environment or in simulated operational environment.
7	**System Prototype Demonstration in an Operational Environment.** Prototype near, or at, planned operational system. Represents a major step up from TRL 6, requiring demonstration of an actual system prototype in an operational environment such as an aircraft, vehicle, or space. Examples include testing the prototype in a test bed aircraft.
8	**Actual System Completed and Qualified Through Test and Demonstration.** Technology has been proven to work in its final form and under expected conditions. In almost all cases, this TRL represents the end of true development. Examples include developmental test and evaluation of the system in its intended weapon system to determine if it meets design specifications.
9	**Actual System Proven Through Successful Mission Operations.** Actual application of the technology in its final form and under mission conditions, such as those encountered in operational test and evaluation. Examples include using the system under operational mission conditions.

other (see Figure 1) shows that the scales are all roughly equivalent and use similar definitions.

Development Phases and Decision Gates

Defining the meaning of the various TRLs is only part of the story, and the real usefulness of employing a TRL system is how it can be used to control the research and development (R&D) process

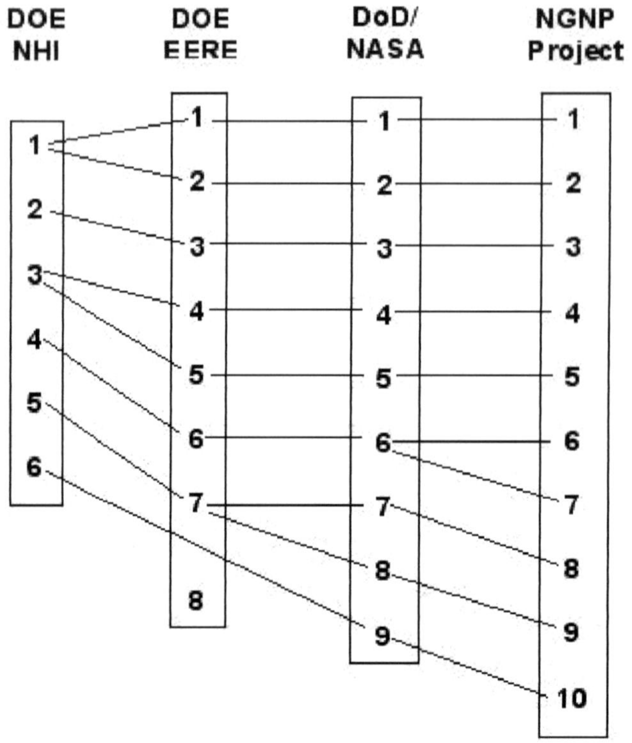

Figure 1: Equivalence mapping of various TRL scales.

and drive development activities. A TRL system is organized as a series of development phases and decision gates, as shown in Figure 2.

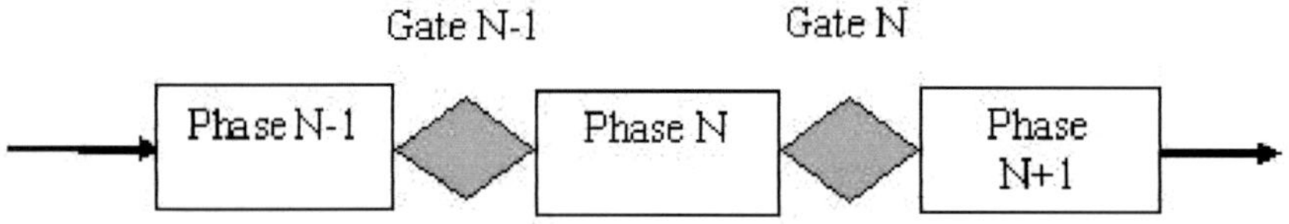

Figure 2: Phase-gate progression for TRL systems.

A development phase is an activity that corresponds to the development work that is needed to achieve a particular TRL, and a decision gate is the point at which a component, sub-system, or system is evaluated and compared to the criteria that must be met in order to achieve the gate's particular TRL. A technology must pass successfully through a gate in order to achieve the gate's TRL. For example, a technology rated at TRL N-1 is undergoing Phase N development. Upon completing the Phase N development activities, the technology is evaluated at Gate N and it is determined that all Gate N success criteria have been met. At that point, the TRL of the technology is increased to N, and the technology becomes eligible for Phase N+1 development. Achievement of the success criteria at a decision gate is a necessary but not sufficient condition to enter into the next phase of development. Sufficient funding and resources are also needed to perform the development work, and advancement of the technology must be supported by the project or program management, and the sponsor for the work.

Technology developers perform the development phases, while assigned evaluators (who ideally are independent from the developers) control the decision gates. Gate decisions are also influenced by the governing program or project manager. An evaluator, based on the information provided by the technology developer, may recommend advancement, or may choose different paths. If the information provided by the technology developer is incomplete, the evaluator may recommend further work in the current R&D phase in order to fill in the gaps. The evaluator may also reject the technology altogether if it appears that the success criteria were not met and may never be met, regardless of how much more effort or funding is expended to achieve the success criteria for the gate. These decision options are illustrated in Figure 3.

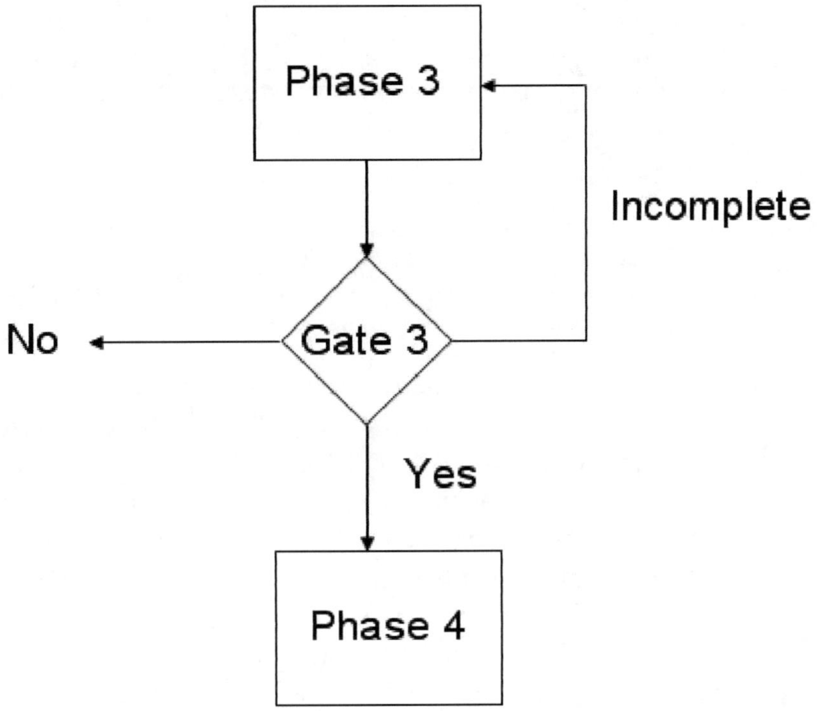

Figure 3: Decision gate options.

Gate decisions are made on a pass-fail basis, and each component, sub-system, or system is evaluated independently of other candidate technologies for a given application. Information collected during the decision gate evaluation process can be used, however, to determine the relative merits of competing technologies, and to determine which of the competing technologies has better performance, lower cost, or greater reliability or robustness. Such decisions are project/program management decisions and fall outside of the scope of the TRL systems described here.

Using a phase-gate process to perform technology development encourages the use of quality assurance tools and the maintenance of development records. Performing a decision gate evaluation will require that information on a particular technology be compiled and assembled in a coherent form and that the information submitted is correct or at least accurate. This in turn will require some level of peer review of the data, which pushes for standardized quality assurance measures for how the work is performed and how the data on a particular technology is recorded and stored. Such information is conveniently stored in a component's or system's case file, which holds all of the records pertaining to a component's or system's development. As a by-product of this process, information to support future intellectual property claims is created almost automatically and can be obtained from the data stored in the case files.

Entry Process

All of the TRL systems described require an entry process. The entry process is a screening process that is used to determine whether an idea, concept, design, or candidate technology may be capable of performing a given task, and that an achievable development path can be envisioned. If development is warrented, a case file is created for the particular technology, and initial screening information is placed into the case file. The entry process is essentially a decision gate, and the process used to perform the evaluation is the same as for the other gates. Unlike the TRL decision gates, the entry gate sits outside of the normal linear progression of the TRL phases and gates and can feed a candidate technology into any of the development phases, as appropriate to the initially assessed level of technology readiness for a given application.

Risk Reduction

Implied in the TRL systems are the assumptions that greater TRLs correspond to greater reliability, that technologies with higher TRLs function more closely to the desired end-state of the technology, and that the cost of performing a validation test increases as the fidelity of testing begins to approach actual end-use conditions. Reliability is the probability of achieving the same result upon repeated trials. The converse of reliability is the probability of failure, or the probability of not achieving the same result upon repeated trials. Advancement through a decision gate requires the performance of one or more successful tests or experiments where success is defined as an outcome meeting or exceeding selected success criteria, and a failed experiment or test represents a "loss" in economic terms in the development process. A failed experiment or test either results in the need to repeat an experiment or test after additional development work (greater cost), or rejection of the technology (loss of investment). Given the directional preference of decision gates ("only successful experiments or tests lead to advancement"), it is better from an investment standpoint to perform less expensive experiments or tests at lower development phases because the cumulative penalty (number of failed experiments times the cost per experiment) may be reduced. At higher TRLs, reliability is greater, so more realistic (and presumably more expensive) tests can be performed with less risk of incurring greater costs or losing the investment. The relationships between investment costs, probabilities of failure, and TRLs are shown qualitatively in Figure 4.

In some cases, costs savings may be achieved by skipping one or more development phases. This may be preferred when the cost of a higher fidelity test is lower than the cost of a lower fidelity test, and the impact of a failure on other related components or systems is minimal. For example, if a large-scale pilot plant with the capacity to test components exists and is currently in operation, then it may be more practical to perform a component test in the actual pilot plant than it would be to build a dedicated test rig and perform experiments on the component in the laboratory prior to testing the component in the pilot plant.

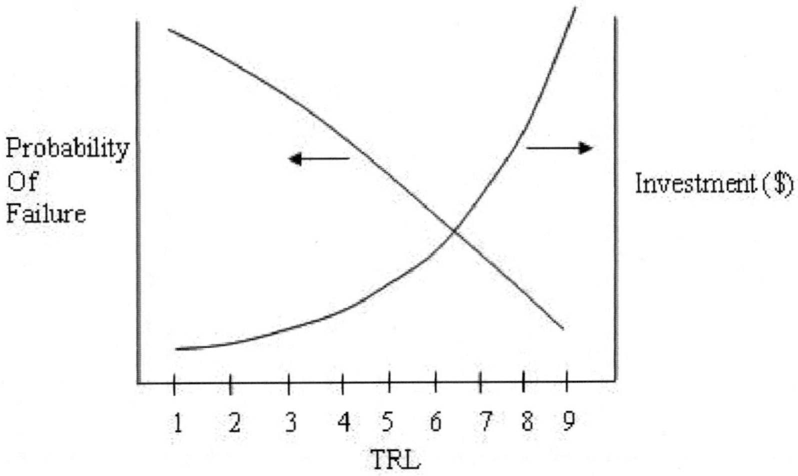

Figure 4: Relationships between investment cost, TRL, and probability of failure.

Management and Communication Tool

Using a TRL system to manage larger technology development projects, regardless of the number of TRLs used or the exact definition of what each TRL means to the development program, can be an organizing framework around which R&D is conducted. Researchers, developers, project/program managers, and program sponsors agree upon a given TRL system, and assume its methods and definitions as a common way of doing business. The TRLs and the development phases then can become a shorthand language that enables people of different technical backgrounds to communicate without getting lost in the details of any particular technology. For the researchers, the requirements for achieving a particular technology readiness level can be translated into technical performance criteria and supporting experiments, while managers and sponsors can interpret the same TRL in the context of project risk, costs, and impact of development activities on a development program's schedule and budget. The progressive nature of the TRL systems establishes a development direction (lower to higher technological maturity) that can be understood by all, and allows for the comparison of competing technologies on a common basis.

Adopting a TRL system separates the development process from the development activities, and this allows for more management control and better definition of what activities and information are needed to achieve higher technology readiness. It divides development work into distinct phases that can be handled in discrete units, which aids in management of the development program. Without explicitly established technology targets, research and development programs may be subject to scope drift, perpetual development, and lack of a clear message to program sponsors. Use of a TRL system to manage development helps prevent these problems, and will help to provide assurance to sponsors and stakeholders that development activities are purposeful and are directed towards achieving specific goals.

References

[1] Mankins, J.C. (April 6, 1995), "Technical Readiness Levels," White Paper, Advanced Concepts Office, Office of Space Access and Technology, National Aeronautics and Space Administration (NASA).

[2] National Aeronautics and Space Administration (2007), *NASA Systems Engineering Handbook*, NASA/SP-2007-6105, Rev. 1.

[3] Nolte, W.L., Kennedy, B.C., and Dziegel Jr., R.J. (2003), "Technology Readiness Calculator," White Paper, AFRL/SNOL, http://www.dtic.mil/ndia/2003systems/nolte2.pdf.

[4] U.S. Department of Energy, Office of Environmental Management (March 2008), *Technology Readiness Assessment (TRA)/Technology Maturation Plan (TMP) Process Guide.*

[5] Sherman, S.R., Wilson, D.F., Pawel, S. (September 2007), "NHI Technical Readiness Evaluation System," *Proceedings of Global 2007 – ANS Conference on Advanced Fuel Cycles and Systems*, Boise, Idaho.

694c Economical Analysis of Ethanol Steam Reformer for Hydrogen Production Associated with a Sugarcane Industry

Antonio Carlos Caetano Souza and José Luz Silveira. Energy, UNESP (São Paulo State University), Av Ariberto Pereira da Cunha, 333, Guaratinguetá, 12514-410, Brazil

An expressive amount of produced hydrogen is generated by customers "in-situ" such as petrochemical, fertilizer and sugarcane industries. However, the most utilized feedstock is natural gas, a non-renewable and fossil fuel.

The introduction of biohydrogen production process associated in a sugarcane industry is an alternative to diminish emissions and contribute to create a CO2 cycle, where the plants capture this gas by photosynthesis process and produces sucrose for ethanol production. The cost of production of ethanol has dramatically decreased (from about US$ 700/m3 in 1970s to US$ 200/m3 today), becoming this a good option at near term, inclusively for its utilization by customers localized in main regions (localized especially in regions such as Southeastern Brazil)

Also in near future, it will possible the utilization of fuel cells as form of distributed generation. Its utilization could occur specially in peak hours, diminishing the cost of investments in newer transmission systems.

A technical and economic analysis of steam reformer of ethanol to hydrogen production associated with sugarcane industry was recently performed. This technique will also allow the use of ethanol when its price is relatively low. This study was based on a previous R&D study (sponsored by CEMIG - State of Minas Gerais Electricity Company) where thermodynamic and economic analyses were developed, based in the development of two ethanol steam reformers prototypes.x

In this study an analysis was performed considering the use of bagasse as source of heat in the steam reforming process. Its use could to diminish the costs of hydrogen production, especially at large scale, obtaining cost-competitive production and permitting that sugarcane industry produces hydrogen in large scale beyond ethylic alcohol, anhydrous alcohol (or ethanol) and sugar.

694d Thermodynamic Evaluation of Processes for Hydrogen Production from Coal – Theoretical and Experimental Studies

Kanchan Mondal and Tomasz Wiltowski. Department of Mechanical Engineering and Energy Processes, Southern Illinois University, Carbondale, IL 62901

The paper presents a procedure for the comparison of various technologies for the production of hydrogen from coal based on First and Second Law Analysis. All known technologies listed in publicly available literature are analyzed based on calculations of the energetic and exergetic analysis of the whole process as well as the sub processes. Energy and exergy losses associated with generation of syngas from coal, separation of hydrogen from COx as well as exergetic loss of emissions are presented. The optimal conditions for each process by minimizing these losses are presented. Results from a flexible lab scale modular experimental bench are presented as experimental verification of the trends in efficiency improvements and for the verification of the theoretical comparisons. Based on the results, the most energy efficient process for a given hydrogen purity is suggested along with the optimal conditions for operation.

694e Process Integration and Optimization of Hydrogen Production Strategies

Wei Yuan, Susilpa Bommareddy, Norman E. Sammons Jr., and Mario R. Eden. Department of Chemical Engineering, Auburn University, 230 Ross Hall, Auburn University, AL 36849-5127

Experimental researchers in the Consortium for Fossil Fuel Science (CFFS) are developing novel technologies for hydrogen production and storage, however the economic and environmental viability of the technologies are mostly evaluated separately rather than in an integrated manner. In this work, we aim to address this gap through the development of integrated system engineering strategies for modeling, integration and optimization of hydrogen polygeneration plants.

We are developing process models for a variety of hydrogen production schemes. Using process integration techniques and advanced computer-aided tools, the systems have been optimized and the economic potential as well as the environmental impact of the technologies evaluated. Several reforming techniques including four developed within CFFS have been studied based on small and large scale production of hydrogen. Literature data along with data obtained in other research projects at Auburn University has been used to develop models for partial oxidation (POX), steam (SR) and autothermal reforming (ATR) of a variety of hydrocarbon resources. In addition, data provided by other researchers in CFFS has been used to build similar models for super critical water reforming (SCWR) of methanol and biomass (using glucose as a model molecule), dry reforming (DR) of methane and finally a catalytic dehydrogenation (CDH) of methane. The last process is a single step process that in addition to high purity hydrogen also produces a valuable carbon nanotube byproduct.

Our work has made it clear, that technologies like POX and ATR due to their exothermic nature are much less sensitive to changes in heating utility cost like the recent spike in natural gas prices. However it is worth noting that all processes can be improved immensely by implementing process integration. An economic analysis of all the generated case studies was performed to evaluate the hydrogen production cost. It was found that the current technologies for producing hydrogen from liquid fuels are not attractive if evaluated only on the production cost. Benefits such as transportability etc. will need to be quantified for all types of fuel in order to better compare the technologies. The environmental impact for each process is evaluated and compared. The results show that SCWR of biomass is most environmental friendly compared to the other processes. Several technologies as being developed by CFFS researchers appear to be potentially cheaper alternatives to the current state of the art.

AUTHOR INDEX

AUTHOR INDEX

American Institute of Chemical Engineers
3 Park Avenue
New York, New York 10016-5991

CURRAN ASSOCIATES INC
proceedings
.com

11th European Conference on Applied Superconductivity

(EUCAS 2013)

Journal of Physics: Conference Series Volume 507

Genoa, Italy
15-19 September 2013

ISBN: 978-1-63439-287-7
ISSN: 1742-6588

IOP Institute of Physics